FUEL
CELLS:

Grove Anniversary Symposium '89

FUEL CELLS:

Grove Anniversary Symposium '89

Proceedings of the
Grove Anniversary
Fuel Cell Symposium,
London,
September 18-21, 1989

EDITOR: D.G. LOVERING

Reprinted from the
Journal of Power Sources
Vol. 29, Nos. 1/2

ELSEVIER APPLIED SCIENCE
LONDON AND NEW YORK

ELSEVIER APPLIED SCIENCE PUBLISHERS LTD
Crown House, Linton Road, Barking, Essex IG11 8JU, England

Sole Distributors in the USA and Canada
ELSEVIER SCIENCE PUBLISHING CO., INC.
655 Avenue of the Americas, New York, NY 10010, USA

British Library Cataloguing in Publication Data

Lovering, David G.
  Fuel cells.
  1. Fuel cells
  I. Lovering, David G. II Journal of Power Sources
  621.31'2429

ISBN 1-85166-816-0

Library of Congress Cataloging-in-Publication Data

Fuel cells : Grove anniversary symposium '89 / editor, D.G. Lovering.
     p.   cm.
   "Journal of power sources, vol. 29 (1990), nos. 1-2."
   Proceedings of the Grove Anniversary Fuel Cell Symposium held at
the Royal Institution, London, 18-21 Sept. 1989.
   Includes bibliographical references.
   ISBN 1-85166-816-0 (U.S.)
   1. Fuel cells--Congresses.   I. Lovering, D. G. (David G.)
II. Grove Anniversary Fuel Cell Symposium (1989 : Royal Institution,
London)
TK2931.F833   1990
621.31'2429--dc20   89-25845
                                                            CIP

Printed in The Netherlands by Krips Repro B.V.

Conference Steering Committee

Chairman: G. J. K. Acres (Johnson Matthey plc)

Technical Secretary: D. G. Lovering (The David Graham Consultancy)

Committee:

D. S. Cameron (Johnson Matthey plc)
J. Campbell (Cookson Group plc)
R. J. Carpenter (British Gas plc)
R. M. Dell (Harwell Laboratory)
S. Fleet (Department of Energy)
S. M. Goodall (IBC Technical Services Ltd)
M. N. Mahmood (BP Research)
R. Parsons (University of Southampton)
D. Pollard (GEC Alsthom)
N. M. Sammes (ICI plc)
B. C. H. Steele (Imperial College)

Session Chairmen

G. J. K. Acres (Johnson Matthey plc)
R. M. Dell (Harwell Laboratory)
D. G. Lovering (The David Graham Consultancy)
R. Parsons (University of Southampton)
B. C. H. Steele (Imperial College)

Sponsoring Companies

British Gas plc
British Petroleum
Cookson Group plc
GEC Alsthom
Imperial Chemical Industries plc
Johnson Matthey plc

Organised by IBC Technical Services Ltd at the Royal Institution, London, U.K.

Contents

Preface

Although William Robert Grove, later Sir William, first expounded the fuel cell principle in a postscript to a letter in *Philosophical Magazine* one hundred and fifty years ago, fuel cells as electricity generators still, unfortunately, remain full of promise for the future.

Notwithstanding, the pace of development for most of these attractive devices has steadily accelerated during the last two decades, as evidenced by the intense interest, fervent activity and large attendances at recent meetings. Thus it was that an *ad hoc* committee of British industrialists, researchers and others were prompted to convene this Symposium to pay tribute to the vision of Grove; it was held at The Royal Institution, London, 18 - 21 September 1989, where he had collaborated with Faraday and many other distinguished scientists of the day. The main purpose was to provide a forum for appraising contemporary developments in fuel cells, their potential role in an increasingly environmentally-aware community and how they might influence future energy strategies, particularly in the United Kingdom.

Well over two hundred delegates from many disciplines and from all around the world, met on this occasion to hear twenty papers, three panel discussions and six short contributions given by international authorities in the field. Apart from two of the latter, all are collected within the present volume and will serve as a reference to events, opinions and views on fuel cells at this time.

The meeting was opened by HRH The Duke of Kent as President of The Royal Institution. Professor J. M. Thomas, its Director, presented the Grove Anniversary Fuel Cell Lecture, amply illustrated by archival material written by Grove; this was followed by spectacular demonstrations of the explosive oxidation of hydrogen in air and concluded with the cool, controlled combination of hydrogen and oxygen in a Ballard solid polymer fuel cell.

The technical sessions of the programme focussed on the many aspects of fuel cells, including their promise to reduce the topical 'greenhouse effect' due to their efficiency, present national development programmes, progress in the several fuel cell types, market assessments, opportunities and prospects through to applications as static generators and mobile power sources.

Speakers identified the present problem areas in the technology, with cost reduction the most prominent. This editor's contention is that fuel cell introductions face the same paradox which surrounds all new technologies in that the benefit of mass production with its implicit cost reduction depends on market size! Other major considerations for fuel cells are lifetimes/ reliability and materials development. Appleby highlighted the latter aspect in his closing remarks, having opened proceedings two days earlier with a comprehensive historical account of the field. We are privileged to include both of his papers here. All authors are to be thanked for their prompt provision of high-quality manuscripts which have ensured rapid publication. A complete record of the proceedings was tape recorded, including panel discussions, and is presently held by this editor.

Overall, the Symposium proceeded smoothly if somewhat hectically. But like all such meetings, it provided that essential opportunity to renew old and begin new acquaintanceships between the *cognoscenti*. International collaboration across the scientific, technical and engineering disciplines will prove to be the *sine qua non* for successful fuel cell implementation in the cleaner, quieter and more efficient world we are all seeking.

DAVID G. LOVERING
Proceedings Editor and Technical Secretary,
The Grove Anniversary Fuel Cell Symposium Committee,
September 1989

Background and Context of Fuel Cell Applications

Journal of Power Sources, 29 (1990) 3 - 11   3

FROM SIR WILLIAM GROVE TO TODAY: FUEL CELLS AND THE FUTURE

A. J. APPLEBY

Center for Electrochemical Systems and Hydrogen Research, TEES/Texas A&M University, College Station, TX 77843 (U.S.A.)

Introduction

The history of the fuel cell dates from Sir William Grove's invention of the gaseous voltaic cell, which he first described in 1839. Grove was the quintessential nineteenth century amateur scientist, who nonetheless made a number of important discoveries. In later editions of his book [1], he describes fuel cells operating on many different reactants, including ethylene and carbon monoxide, as well as hydrogen. He should also be remembered for being perhaps the first person to appreciate the law of the conservation of energy [2], anticipating Helmholtz by one year. His scientific discoveries are all the more remarkable for the fact that his public recognition was as a jurist.

Grove published a description of the working of the first fuel cell in February 1839 [3]. This included the mechanism of a single cell, consisting of hydrogen and oxygen in contact with two platinized platinum electrodes. In that paper, he alluded to the possibility of combining several of these in series to form a gaseous voltaic battery, which he described for the first time in 1842 [4]. This description was followed by other papers [5].

The battery consisted of fifty single cells, each with anodic and cathodic platinized platinum foils one quarter of an inch wide. The most important observation that Grove made was his famous statement concerning the necessity for a 'notable surface of action' between the gas, electrolyte and electrode phases in his cells. His language makes many of the points that have been reiterated, in somewhat different form, during the age of the modern fuel cell, since about 1955.

Grove's words in his 1842 paper are worth quoting again:

"As the chemical or *catalytic* action . . . could only be supposed to take place, with ordinary platina foil, at the line or water-mark where the liquid, gas and platina met, the chief difficulty was to obtain anything like *a notable surface of action.* I determined to try the platina platinized .... It is obvious that, by allowing the platina to touch the liquid the latter *would spread over its surface by capillary action and expose an extended superficies to the gaseous atmosphere.*"

The expressions in italics, taken together, constitute the leitmotif of the development of today's fuel cell electrodes.

0378-7753/90/$3.50

As he stated himself, Grove's series of fuel cells were hardly practical devices for power production from hydrogen and oxygen, indeed, they were scarcely more than capable of parlor demonstrations. Their capabilities for delivering current were strictly limited by the small effective active area of each electrode, which was probably little more than 10 mm^2, representing a single meniscus about 2 mm high on a piece of platinized platinum sheet 6 mm wide. However, as the above quotation shows, he did realize the need for the highest area of contact between the electrolyte, the gaseous reagent and the electrocatalytic conductor, *i.e.*, the 'notable surface of action'. Trying to acquire this optimized reaction surface has remained the basis of fuel cell research and development ever since. Because of this realization, Grove can be truly said to be the inventor of the fuel cell.

The first modern fuel cell structure: Mond and Langer

Following Grove, the concept of increasing the 'surface of action' as a means of increasing performance was taken up by Mond and Langer [6], whose June 1889 paper gives a list of some fifteen papers published since Grove's first experiments, including one by Lord Rayleigh [7]. However, they state that up to that time the subject has been given 'very little attention'.

Mond and Langer were the first workers to try to improve upon Grove's electrodes by giving them a three-dimensional form. Grove's electrodes had a two-dimensional meniscus in which current was collected parallel to their plane. Mond and Langer made the electrode structure porous, and rotated it by 90°, thus creating a structure with all the features of the modern fuel cell. Each consisted of a diaphragm, made of a porous non-conducting substance (plaster of Paris, earthenware, asbestos, pasteboard), with electrodes consisting of perforated platinum or gold leaf as a current collector, contacting active surfaces of platinum black. The diaphragms contacted the gases on each side, and could be placed 'side by side or one above the other'. Their cells operated on hydrogen and oxygen at 0.73 V and a current density of 3.5 mA/cm^2. In contrast, the phosphoric acid fuel cell of today operates at the same voltage, but at a current density 60 times higher, and more advanced systems can increase this current density by a further factor of ten. Apart from using newer materials, these cells are fundamentally similar to the Mond and Langer design, and they use a thin diaphragm carrying the electrolyte to reduce internal electrical resistance, always a major irreversible loss at high current density.

The major technological change in modern cells is a microengineered control over the meniscus, which in the Mond and Langer cell would have flooded the electrode structure, thus reducing the internal volume open to the gaseous reagent, the area available for reaction, and thus the current density. For this reason, Mond and Langer's cells showed a voltage that decreased by about 10% per hour as a function of time, as product water collected in the acid electrolyte at the oxygen cathode. In modern cells, a great increase in the 'surface of action' with prevention of flooding is carried out by two approaches: the use of graded porosities, so that suitable micro-interfaces are maintained by capillary action, or by the use of a non-wettable additive, which locally creates zones empty of electrolyte, serving as gas microchannels within the electrode structure. These are discussed later.

A further advance was Mond and Langer's realization of the great efficiency of the electrochemical process compared with that of the thermo-dynamic engine. "We prefer to work ... with an e.m.f. of about 0.73 V ..., which gives a useful effect of nearly 50% of the total energy contained in the hydrogen absorbed in the battery." They also showed that water gas containing 30 - 40% hydrogen, produced by the gasification of coal would operate the fuel cell, at least for short times.

In summary, Mond and Langer's concept contained all the elements of the modern low-temperature fuel cell, except a means for maintaining an optimized three-dimensional 'notable surface of action' within the electrode film, instead of the less efficient two-dimensional structure used by Grove. Though it may not have been widely noticed, the year of this Symposium is therefore also the centennial of the first modern fuel cell concept.

The age of coal

Mond and Langer had realized that coal could be used as a source of hydrogen for the fuel cell, whereas Grove stressed only the use of pure hydrogen derived, for example, from zinc dissolution. While coal would have been used to produce the zinc, the overall concept would have had an energy efficiency too low to be of practical value. In work following that of Mond and Langer, emphasis shifted from hydrogen to more common and practical fuels such as coal, reflecting Ostwald's visionary hope of 1894 that the 20th Century would become the Age of Electrochemical Combustion, with the replacement of the steam Rankine cycle heat engine by much more efficient, pollution-free, fuel cells. In his well-known paper [8], Ostwald stressed the wastefulness of the steam-engine, with its then heat-to-work efficiency of 10%. He stated that the new way must be founded on electrochemistry, which could allow the theoretical amount of work to be obtained from coal, acid electrolyte and air, so that fire would not be the only method of effecting change, but would in future be replaced by electricity. One aspect, familiar to modern ears, that he stressed would be the lack of pollution: 'kein Rauch, kein Russ', no smoke, no soot. The dream of the late nineteenth century remains the dream of the late twentieth.

Ostwald's vision did not happen, for reasons that were largely connected with the slow electrochemical reaction rates of common fossil fuels, but which also resulted from the advent of the various types of internal combustion engine using cleaner liquid (or gaseous) fuels. Competition from the latter resulted in the demise of the electric storage battery for transportation applications, and led to a lack of interest in electrochemical power in general, and in the development of the fuel cell in particular. We can also suppose that this Utopian view of the world, as seen from the perspective of the nineteenth century, might have become closer to reality had it not been for the wars of the twentieth, and the attendant cynicism and lack of care for human progress that they have characterized.

The outstanding researchers of the 'Age of Coal' were Jacques [9] and Baur [10]. Jacques, working about 1895, seems to have been the first to make large systems, including a 1.5 kW battery with about 100 small tubular cells, and later, a battery with a design power of about 30 kW. The cells consisted of cylindrical iron pots as cathodes, supplied with air via distribution tubes, with internal coke rods (3 in the case of the 1 kW-scale cells) as cathodes. Unlike Mond and Langer, with their modern concept of the immobilized electrolyte for mechanical control, Jacques used free electrolyte cells with molten KOH at about 450 °C. High performance (reputedly 100 mA/cm^2, 1.0 V) was obtained, so that the large cells (about 1.2 m high, 0.3 m diameter) were capable of over 300 W each. Such coke–air cells were suggested for all-electric naval vessels, as well as for urban power: 'think of a smokeless London', is a quotation from his article.

Jacques' claims were investigated by Haber and Brunner [11] in 1904, who showed that the anode operated via redox systems (manganite–manganate) giving favorable kinetics. They showed that Jacques' electrolyte was not invariant, since carbonate formation resulted as the carbon anode reacted. Today we realize why the iron cathode in the molten caustic electrolyte does not require a 'notable surface of action'. It is also a redox system involving molecular oxygen chemically dissolved as peroxide, so long as the electrolyte remains as hydroxide. After the rapid neutralization of the electrolyte, reaction stops. The spent electrolyte must be replaced or regenerated, in practice probably requiring more energy than the cell produces. Jacques thought that the nitrogen in the air bubbling through the melt would be sufficient to remove the excess carbonate as carbon dioxide, but unfortunately this is not so. While he acknowledged that his cell was not suitable for use with coal, because of its lack of adequate conductivity and to the large amount of ash produced (he did not mention sulfur), it was clear that KOH was not a practical electrolyte for use with carbon. The cell was essentially a large primary battery, unsuitable for continuous use.

Baur and Ehrenberg [10] attempted to use coal directly, again using carbon as a model, but this time with invariant electrolytes, such as molten carbonates. After this time, interest fell in the direct use of carbon (coal) as a fuel. The problems remained the same: the formation of ash, the poor conductivity, and the need for some method of continuously feeding the solid fuel. Muscle-power was of course economically adequate for this in the nineteenth century.

Improvements

As can be seen from the above, fuel cell developments from Grove's time to the first half of this century resulted in a series of dead-ends. Progress, or rather lack of it, revolved around the use of unsuitable chemistry and faulty engineering. While it is easy to speculate, with hindsight, on what might have been, the elements to make practical devices were there. Baur, *et al.* [12], working in 1922, used molten carbonate electrolyte with gas-operated anodes. In many respects, their cell structure was a high-temperature version of that of Mond and Langer, though with tubular, rather than planar, geometry. For the first time, they conciously used the concept for maintaining the 'notable surface of action' of the electrodes, already implicit in Mond and Langer's work, by the use of capillarity. The electrolyte of the cell was contained in a matrix powder material corresponding to Mond and Langer's diaphragm. If this is in contact with an electrode in porous form, whose range of pore diameter overlaps that of the matrix, capillarity will determine that if the latter is filled with electrolyte that wets all the components, then fine pores in the electrode will be filled with electrolyte, whereas coarse pores will remain empty. Thus, the latter will be filled with gas, and the resulting convoluted structure will supply the required 'surface of action'.

Starting in 1933 with the earliest work of Bacon in England (reviewed in ref. 13), the fuel cell can be said to have reached adolescence. Bacon wished to use ordinary materials, *i.e.*, no noble metals, a non-corrosive environment for maximum lifetime, and the highest possible electrode reaction rates (measured in current density, A/cm^2) at the highest practical cell voltage, *i.e.*, efficiency. Reaction rates increase at high temperatures and pressures, therefore Bacon, associated with Parsons' turbines, marine engineering and high-pressure boilers, used the engineer's approach. His fuel cell was essentially in a high-pressure boiler. While he would have preferred to use steel, nickel was the best compromise. It and its oxide are stable in alkaline solution at both the hydrogen and oxygen electrode, respectively, though it is not stable in acid. Hence, the electrolyte was hot potassium hydroxide solution, circulated to remove heat from the high-power system, and to remove water. To maintain an invariant electrolyte composition, a carbonaceous fuel, or air containing carbon dioxide, was excluded. Hydrogen was therefore the fuel, as in Grove's cells, so that with pure oxygen as the complementary oxidant the product was pure water. Bacon conceived his system as a storage unit, in which hydrogen and oxygen could be produced with off-peak electrical power, and consumed when required. He thus had in mind Grove's "effecting the decomposition of water by means of its decomposition... (which) exhibits such a beautiful example of the correlation of natural forces".

Circulation meant a free electrolyte, hence Bacon used an electrode structure that combined the porous properties of Mond and Langer's and Baur, Treadwell and Trumpler's matrix and electrodes: a wetted, fine pore structure facing the electrolyte, and a more open structure on the gas side. In this respect, he followed the earlier work of Schmid [14], who developed the first 'Diffusionsgaselektrode' in 1923 for use in aqueous acid electrolytes. This had a dual-porosity structure with a coarse-pore graphite gas-side layer and a fine-pore platinum electrolyte layer.

Bacon's cell thus combined this with a distillation of the work of previous researchers: pure hydrogen and oxygen from Grove, the parallel structure of Mond and Langer, the electrolyte of Jacques without the error of carbonatation, and a compromise in temperature and composition between the purely aqueous and molten electrolyte systems.

For a long cell lifetime, the temperature of the Bacon cell was limited to 200 °C, which meant that a pressure of 45 atm was possible to achieve high performance. Bacon thus obtained about 1 A/cm^2 at 0.8 V, or 0.4 A at 0.85 V, which would be considered remarkable even today. Parallel with Bacon's later work was that of Justi and Winsel in Germany [15] which achieved a similar performance at lower pressures and temperatures by increasing the nickel electrode internal area, $i.e.$, the 'notable surface of action'. This they did by means of the high-surface area Raney nickel DSK (Doppelskelett Katalysator) electrode, whereas Bacon used electrodes made from sintered carbonyl nickel powders. This reflects the fact that electrode structures give us a choice: high temperature increases activity, but causes loss of surface area by sintering and compaction, therefore all structures and operating conditions are compromises.

Bacon continued to develop his cell up to the early 1960s, as long as funding was available in England. After this, the concept was transferred to the Pratt and Whitney Division of United Aircraft Corporation (now United Technologies Corporation) in Connecticut, where it was modified for space use. This required reduction of the pressure vessel weight by reducing the pressure by a factor of ten, which required a simultaneous increase in temperature to 260 °C and an increase in electrolyte concentration to 75% KOH to prevent boiling. Similarly, the circulating electrolyte was eliminated, and heat and water were removed by a closed-loop hydrogen cycle. The increased temperature did not compensate for the lowered pressure from the performance viewpoint, but the system was still capable of 0.15 A/cm^2 at 0.85 V [16].

Bacon's cell, as modified by Pratt and Whitney, was the on-board power system for the Apollo lunar missions. Without this technology, the only one available at the time with sufficient power and energy densities, the lunar landings would have been impossible.

The non-wetting electrode

Above, we alluded to another method by which Grove's 'notable surface of action' could be achieved: the notion of 'controlled wetting'. While this is almost inconceivable in a molten electrolyte, in aqueous solutions it is possible because of the lyophobic and lyophilic properties of many materials. By the early 1930s, the need for a new concept to open up the electrode porosity was realized. One method, that of Tobler [17] in 1933, made the electrode into a thick stationary bed with much open porosity. This was not satisfactory, because it separated the active parts of the electrodes by too great a distance, and thus introduced a large internal resistance. Mond and Langer's 'closest approach' concept for the anode and cathode could thus not be used. Wetproofing electrodes (with highly unsuitable paraffin wax) was perhaps first used by Heise and Schumacher in 1932 [18].

After about 1950 Teflon (ptfe) became available. It was first used in platinum electrodes for acid electrolyte, and carbon electrodes for alkaline electrolyte, before 1964 - 1965 at General Electric and Union Carbide respectively [19, 20]. The use of this remarkable material made the aqueous (liquid) electrolyte fuel cell in its modern form possible.

New developments of the 1960s

In the early 1960s, attention turned again to the platinum-catalyzed acid electrolyte cell in two different forms. One used a polymer acid electrolyte, which made it simple and reliable. Its combined electrode–electrolyte structure made it automatically water-rejecting, and at its original modest power levels (37 mA/cm^2), it required no wetproofing. Its original electrolyte material restricted its operating temperature and thus its performance, but it was developed by General Electric for the modest power requirements (1 kW in a 29 kg unit) of the Gemini missions, where its ability to produce potable water for the astronauts was a great advantage in the lightweight capsule. Its test vehicle, a small General Electric fuel cell, became the first to go into space in a suborbital flight on October 30, 1960.

The second acid technology was developed to attempt to use carbonaceous fuels directly, which is impossible in cells with alkaline electrolyte. The chosen electrolyte made use of the great stability of phosphoric acid to obtain the highest operating temperature possible (150 - 200 °C) for the greatest reaction rates. The breakthrough was the use of stable ptfe as a wet-proofing agent in the high-area platinum black electrodes. These allowed the whole inside area of the electrode to become a convoluted meniscus, increasing the 'surface of action' to close to its physical limits. However, carbonaceous fuels still showed disappointingly low rates in this fuel cell environment, even with excessive amounts of noble metal catalysts at the anode. These low rates were due to electrocatalytic poisoning effects. The above work is reviewed in ref. 21. As other speakers at this Symposium describe, the phosphoric acid fuel cell is now favored for utility use (stationary power generation), where its waste heat at almost 200 °C is a valuable source of energy for raising steam to produce hydrogen mixtures by reforming of natural gas.

In parallel, work started in the Netherlands under Broers and Ketelaar on molten salt fuel cells using molten carbonate electrolyte at 650 °C [22].

These followed earlier work of Greger [23] and Gorin [24], which were in turn based on that of Baur *et al.* [12]. The molten carbonate cells could use nickel-based electrodes made from sintered powder similar to Bacon's. A mixed alkali metal carbonate was the only molten salt allowing low-polarization electrode reactions with carbonaceous fuels. Finally, Weissbart and Ruka [25] raised the operating temperature further, to 1000 °C, and adapted the doped zirconia conducting ceramic oxide of the 'Nernst Glower' of 1900 [26] as a solid electrolyte.

From the beginning, it was noted that laboratory high-temperature cells would not operate directly on hydrocarbons, which showed cracking and, as in the medium-temperature phosphoric acid cell, disappointing reaction rates. It was finally shown that reformate (the gas resulting from steam-reforming of the hydrocarbons) was an effective fuel, which could be used directly in the high-temperature cells, and after water-gas shifting to prevent poisoning by carbon monoxide in the phosphoric acid system. Thus, hydrogen (in the form of mixtures) had again become the fuel of choice. By the late 1960s, these developments had laid the groundwork for the fuel cell developments of the last twenty years.

Today and the future

The fuel cells developed since 1970 have been characterized by the elimination of diffusion limitations in electrode structures by better understanding of the nature of Grove's 'surface of action', by the reduction of costly catalyst loadings in the phosphoric acid system by a factor of about 200, still accompanied by an increase in performance, and finally, improvements in lifetime, making practical operation over five years or more possible. One disappointment has been the lack of breakthroughs in increasing the specific rates of the oxygen electrode process in low- and medium-temperature systems, though some progress has been made.

Above all, emphasis has been on engineering. When hydrocarbon feedstock (*e.g.*, methane) is used as fuel, the chemical plant converting it to impure hydrogen must be integrated with the fuel cell to ensure maximum efficiency. Now, promised efficiencies based on the higher heating value of methane at practical electrode reaction rates have be raised to about 45% in phosphoric acid systems and to 50% and 60% in the solid oxide and molten carbonate systems respectively. These high efficiency values, combined with very low atmospheric emissions of fuel cell systems and the possibility of dispersed, modular units, may soon revolutionize power generation, one century after Ostwald's prediction. They may come in time to have some impact on the 'greenhouse effect', and tropospheric pollution in general.

Specific power outputs of fuel cell generators have increased dramatically in twenty years, in the case of the pure hydrogen–alkaline aerospace system by two orders of magnitude. If pure hydrogen is used as the fuel, the large chemical engineering plant associated with the use of carbonaceous fuels can be eliminated, allowing the use of lightweight non-polluting fuel cells with 50 - 60% thermal efficiencies in transportation. In the future, pure hydrogen may therefore be the fuel of choice, as Grove envisaged.

References

1 W. R. Grove, *The Correlation of Physical Forces*, Longmans Green, London, 6th edn., 1874.
2 W. R. Grove, *The Correlation of Physical Forces*, Longmans Green, London, 1846.
3 W. R. Grove, *Phil. Mag., Ser. 3, 14* (1839) 127.
4 W. R. Grove, *Phil. Mag., Ser. 3, 21* (1843) 417.
5 W. R. Grove, *Proc. R. Soc. London, 4* (1843) 463; *5* (1845) 557.
6 L. Mond and C. Langer, *Proc. R. Soc. London, 46* (1889) 296.
7 Lord Rayleigh, *Proc. Cambridge Phil. Soc., 4* (1882) 198.
8 W. Ostwald, *Z. Elektrochem., 1* (1894) 122.
9 W. W. Jacques, *Harper's Magazine 94* (Dec. 1896 - March 1897) 144.
10 E. Baur, *Z. Elektrochem., 16* (1910) 300; E. Baur and H. Ehrenberg, *Z. Elektrochem., 18* (1912) 1002; E. Baur, A. Petersen and G. Fullemann, *Z. Elektrochem., 22* (1916) 409; E. Baur and J. Tobler, *Z. Elektrochem., 39* (1933) 169.
11 F. Haber and L. Bruner, *Z. Elektrochem., 10* (1904) 697.
12 E. Baur, W. D. Treadwell and G. Trumpler, *Z. Elektrochem., 27* (1921) 199.
13 A. M. Adams, F. T. Bacon and R. G. H. Watson, in W. Mitchell, Jr. (ed.), *Fuel Cells*, Academic Press, New York, 1963, p. 129.
14 A. Schmid, *Die Diffusionsgaselektrode*, Enke, Stuttgart, 1923.
15 E. W. Justi and A. W. Winsel, *Cold Combustion Fuel Cells*, Steiner, Wiesbaden, 1962.
16 B. J. Crowe, Fuel cells: a survey, *Report SP-5115*, NASA, Washington, DC, 1973.
17 J. Tobler, *Z. Elektrochem., 29* (1933) 148.
18 G. W. Heise and E. A. Schumacher, *Trans. Electrochem. Soc., 52* (1932) 383.
19 H. Maget and R. Roethlein, *G. E. Tech. Inf. Serv. Rep. 64DE8*, Lynn, MA, May 1964; L. W. Niedrach and H. R. Alford, *J. Electrochem. Soc., 112* (1965) 117.
20 M. B. Clark, W. G. Darland and K. V. Kordesch, *Proc. 18th Ann. Power Sources Conf., Red Bank, NJ, 1964*, p. 11.
21 H. A. Liebhafsky and E. J. Cairns, *Fuel Cells and Fuel Batteries*, Wiley, New York, 1968.
22 G. H. J. Broers and J. A. A. Ketelaar, in G. H. Young (ed.), *Fuel Cells*, Vol. 1, Rheinhold, New York, 1960, p. 78.
23 H. H. Greger, *U.S. Pat. 2 175 523*.
24 E. Gorin, *U.S. Pats. 2 570 643, 2 581 651, 2 654 661, 2 654 662*.
25 J. Weissbart and R. Ruka, *J. Electrochem. Soc., 109* (1962) 723.
26 W. Nernst and W. Wald, *Z. Elektrochem., 7* (1900) 373.

Journal of Power Sources, 29 (1990) 13 - 28

THE IMPORTANCE OF FUEL CELLS TO ADDRESS THE GLOBAL WARMING PROBLEM

MICHAEL P. WALSH

2800 North Dinwiddie Street, Arlington, VA 22207 (U.S.A.)

Introduction: overview of the problem

During the last decade petroleum supply disruptions and cost increases sent a shock wave through the world which dramatically accelerated interest in more efficient motor vehicles. The need for remedies became a high priority for many countries and investigations were initiated into possible alternative fuels, new technologies and use of highly efficient conventional approaches. As the crises passed, oil prices dropped and interest in these alternatives for energy reasons dwindled; at the same time interest began to increase for environmental reasons. This is especially true for motor vehicles.

Motor vehicles, using petrochemical fuels, emit significant quantities of carbon monoxide, hydrocarbons, nitrogen oxides, fine particles and lead, each of which in sufficient quantities can cause adverse effects on health and the environment. Because of the growing vehicle population and the high emission rates from these vehicles, serious air pollution problems have become an increasingly common phenomenon in modern life. Initially, these problems were most apparent in center cities but recently lakes and streams and even forests have also experienced significant degradation. As more and more evidence of man made impacts on the upper atmosphere accumulates, concerns are increasing that motor vehicles are contributing to global changes which could modify the climate of the entire planet [1 - 8].

In an effort to minimize the motor vehicle pollution problem, emission rates from cars have been limited since the 1980s. Starting in 1975, the pace of control was accelerated with the introduction of catalytic converters on cars in the United States. Initial oxidation catalysts have been replaced by three way converters which can lower carbon monoxide, hydrocarbons and nitrogen oxides simultaneously and increasingly this technology is being applied to vehicles all across the world. Catalytic technology using platinum is now routinely applied to vehicles in Austria, Australia, Canada, Federal Republic of Germany, Japan, Netherlands, South Korea, Sweden, Switzerland, and the United States. Within the next few years, Brazil, Mexico and Taiwan along with most of Europe are scheduled to join their ranks.

The primary impetus for these controls to date has been concerns regarding tropospheric or low level pollution. However, evidence now shows that control of CO, HC and NOx is also important for reducing the risk of global warming.

With regard to carbon dioxide emissions which are also important in this regard, serving as a blanket to trap heat close to the planet, very little progress is occurring. The governmental push of the late 1970s and early 1980s toward improved vehicle fuel efficiency has stalled and market competition now appears to be focused primarily on performance improvements rather than fuel economy gains. Since consumption of each gallon of gasoline results in about 6 pounds of carbon (C) or 22 pounds of CO_2, it is easy to see why CO_2 overall is increasing. As illustrated in Fig. 1, based on projected increases in vehicles and their use around the world, motor vehicle CO_2 emissions will skyrocket over the next forty to fifty years. Modest efficiency improvements on the scale of 1% per year would barely reduce this growth. Most observers would agree that motor vehicles already play a significant role in local, regional and global environmental problems and have the potential to play an even greater role in the future. In addition, vehicles are the major consumer of increasingly scarce oil throughout the world. The purpose of this analysis is to examine their role in these problems and likely future directions. It will show that these problems are directly linked; more vehicles leads to more vehicle miles travelled which leads to more oil consumption and more local and global air pollution. Further more global air pollution especially global warming will exacerbate local and regional pollution problems and vice versa.

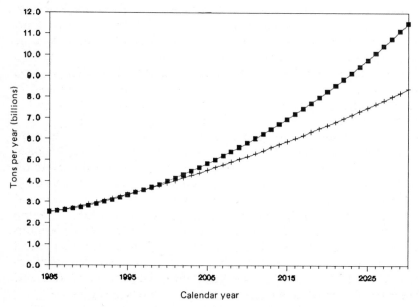

Fig. 1. Global motor vehicle CO_2 emissions, assumed 2% annual VMT growth: ■, constant efficiency; +, 1% annual m.p.g. gain.

First, the next section will assess the important greenhouse gases and the current significance of vehicle emissions. Then, historical and likely future trends in vehicles and their use will be summarized. Finally, the potential benefit from increased use of fuel cells in the transport sector will be summarized.

The role of the motor vehicle in climate modification

Important greenhouse gases

Important greenhouse gases include carbon dioxide (CO_2), CFC 11, CFC 12, methane (CH_4), nitrous oxide (N_2O), ozone (O_3) and the compounds which cause ground level ozone to form, hydrocarbons (HC) and the oxides of nitrogen (NOx). On a global level, each of these gases has been increasing.

As noted in the WMO report [9]:

"The concentrations of halocarbons, methane, nitrous oxide (N_2O), odd nitrogen and carbon monoxide appear to be increasing at present on a global basis, by 5% per year for CFC 11 and CFC 12, 7% per year for CH_2CCl_3, 1 percent a year for CH_4, 0.2% per year for N_2O and 1 to 2% per year for CO."

Concentrations of ground level ozone are increasing, and stratospheric ozone is being destroyed globally. During the Antarctic spring a 'hole' the size of North America is depleted of ozone and, at certain altitudes, is destroyed almost completely because of man made chemicals [10]. Researchers who recently reanalyzed a European data set on tropospheric ozone concentrations from the turn of the century concluded that ozone concentrations had doubled over the past 100 years [11]. One commentator described the finding "as remarkable as the observation of a hole in the stratospheric ozone layer over Antarctica and potentially is just as consequential" [12]. An analysis of several sites indicates that tropospheric ozone background levels are increasing at a rate of 1 to 3% per year with overall NOx increases the controlling factor [13].

Likely effects of climate modification

Over the next fifty years, increasing concentrations of tropospheric ozone and other greenhouse gases are projected to increase the global average temperature between 1.5 and 4.5 °C. Changes likely to accompany this temperature increase include stratospheric cooling; global mean precipitation increase; reduction of sea ice; polar winter surface warming; summer continental dryness; high latitude precipitation increase; and, rise in global mean sea level. Most of these changes should occur gradually (*e.g.*, EPA's recent estimate that average sea levels will rise 5 to 15 inches above current levels by 2025 [14]) if events develop as anticipated; however, the Antarctic ozone hole experience reinforces the anxiety that is associated with any such significant and poorly understood phenomena because of the risk that chemical modifications once initiated may proceed at a faster rate than anticipated.

Carbon monoxide also plays a role

Some of these compounds react with each other in ways only recently understood. For example, hydroxyl radicals (OH) which scavenge many anthropogenic and natural trace gases from the atmosphere, are themselves removed by carbon monoxide [15, 16]. As summarized by Ramanathan recently [17]:

"The highly reactive OH is the primary sink for many tropospheric gases and pollutants including O_3, CH_4, CO, and NO. Hence, increases in CH_4, such as those during the last century [135% increase] could have caused a substantial (20 to 40%) reduction in OH, which in turn, could cause an increase in tropospheric O_3 by as much as 20%. Since CH_4 oxidation leads to the formation of H_2O, an increase in CH_4, an important greenhouse gas, can lead to an increase in H_2O in the stratosphere. Likewise, an increase in the CO concentration can tie up more OH in the oxidation of CO. Thus, through chemical reactions, an increase in either a radiatively active gas such as CH_4 or even a radiatively inactive gas such as CO can increase the concentration of several important greenhouse gases."

Thus carbon monoxide emissions are very important for climate modification. This point was reinforced by MacDonald in a recent analysis:

"Carbon monoxide could thus be indirectly responsible for increasing greenhouse warming by 20 to 40% through raising the levels of methane and ozone. Carbon monoxide participates in the formation of ozone, and also in the destruction of hydroxyl radicals, which are principal sinks for ozone and methane greenhouse gases. Because carbon monoxide reacts rapidly with hydroxyl, increased levels of carbon monoxide will lead to higher regional concentrations of ozone and methane. Measures to reduce carbon monoxide emissions will assist in controlling greenhouse warming." [18]

This is especially significant in view of the evidence that *global* CO levels are now also increasing. As recently noted by Khalil and Rasmussen:

"the average tropospheric concentration of CO is increasing at between 0.8% and 1.4% per year, depending on the method used to estimate the trend, and the 90% confidence limits of the various estimates range between 0.5% and 2.0% per year." [19]

Motor vehicles emit many of these gases

Motor vehicles generate more air pollution than any other *single* human activity. Hothouse gases emitted by (or attributable to) motor vehicles include CFCs, carbon dioxide (CO_2), nitrous oxide (N_2O), methane (CH_4), and the precursors to ground level ozone, hydrocarbons and nitrogen oxides [20].

CFCs. These are the most potent hothouse gases, now contributing about 15 to 20% of the total global warming effect. About 40% of the United States production of CFCs and 30% of European production is devoted to air conditioning and refrigeration. Mobile air conditioning accounted for 56 500 metric tons of CFCs, 28% of the CFCs used for refrigeration in the United States, or about 13% of total production. In contrast, home refrigerators accounted for only 3800 metric tons [21]. Thus, approximately one of every eight pounds of CFCs manufactured in the U.S. is used, and emitted, by motor vehicles. (CFCs are also used as a blowing agent in the production of seating and other foamed products but this is a considerably smaller vehicular use.)

Carbon dioxide. CO_2 is the other major hothouse gas. A single tank of gasoline produces between 300 and 400 pounds of CO_2 when burned. Overall, the transport sector uses approximately 56 quads of energy each year. The consumption varies considerably between regions of the world, with the U.S. far and away the largest consumer, using over 35% of the world's transport energy. Transport consumes almost one thrid of the total energy consumed in the world, again highly variable by region. It is important to note the already significant proportion of energy consumption by transport in many rapidly developing areas of the world.

Because of this high overall energy consumption, it is not surprising that motor vehicles emit over 1100 Tg of carbon, approximately 25% of the world's output [22].

CO, HC and NOx. During 1987, transportation sources were responsible for 40% of U.S. nationwide lead emissions, 70% of the CO, 34% of the volatile organic compounds (HC), 45% of the NOx and 18% of the particulate. In some cities, the mobile source contribution is even higher. Even these contributions do not include the impact of 'running losses', gasoline vapors emitted from the vehicle while it is driving. Accounting for these emissions, the vehicle contribution rises significantly. Including running losses, the overall contribution of transportation to total nationwide HC emissions rises to approximately 50%.

Motor vehicles also dominate the emissions inventories of most European countries, as well. OECD recently noted that,

"The primary source category responsible for most NOx emissions is road transportation roughly between 50 and 70%... Mobile sources, mainly road traffic, produce around 50% of anthropogenic VOC emissions, therefore constituting the largest man made VOC source category in all European OECD countries." [23]

Beyond the U.S. and Europe, Table 1 shows that, for OECD countries as a whole, motor vehicles are the dominant source of carbon monoxide, oxides of nitrogen and hydrocarbons [24].

These pollutants cause other adverse effects

Many of the same pollutants which cause or contribute to global warming, also contribute substantially to adverse health effects in many individuals, in addition to harming terrestrial and aquatic ecosystems, causing crop damage and impairing visibility. Some of these other effects will be described below.

TABLE 1

Motor vehicle share of OECD pollutant emissions (1000 tons, 1980)

Pollutant	Total emissions	Motor vehicle share
NOx	36019	17012 (47%)
HC	33869	13239 (39%)
CO	119148	78227 (66%)

Tropospheric ozone

Photochemical smog results from chemical reactions involving *both* hydrocarbons and nitrogen oxides in the presence of sunlight. While historically the major strategy for reducing smog has focused on tight restrictions on hydrocarbon emissions, NOx control is also necessary. As recently noted by a prominent researcher in this field:

"Recent research results from our research group indicate there is a critical need to consider controls on *both* nitrogen oxides and reactive hydrocarbons if overall oxidant levels are to be lowered... A critical implication of these findings is that without controls on nitrogen oxides the current control policies will simply change the urban ozone problem into a regional scale one." [25]

The ozone problem is a special concern. First, the problem is widespread and pervasive and appears likely to be a long term problem in many areas of the world unless significant further controls are implemented. For example, over 100 million Americans currently reside in areas which exceed the current air quality standard [26, 27]; many of these individuals suffer eye irritation, cough and chest discomfort, headaches, upper respiratory illness, increased asthma attacks, and reduced pulmonary function as a result of this problem.

In addition, the current air quality standard tends to understate the health effects. For example, as noted in testimony before the U.S. Congress in 1987 by EPA, new studies indicate:

"that elevated ozone concentrations occurring on some days during the hot summers in many of our urban areas may reduce lung function, not only for people with preexisting respiratory problems, but even for people in good health. This reduction in lung function may be accompanied by symptomatic effects such as chest pain and shortness of breath. Observed effects from exposures of 1 to 2 h with heavy exercise include measurable reductions in normal lung function in a portion (15 - 30%) of the healthy population that is particularly sensitive to ozone." [26]

Other studies presented at the recent U.S. Dutch Sumposium on ozone indicate that healthy young children suffer adverse effects from exposure to ozone at levels below the current air quality standard [28]. Numerous studies have also demonstrated that photochemical pollutants inflict damage on forest ecosystems and seriously impact the growth of certain crops [29].

It is important to note that global warming may have a significant impact on local ozone air pollution episodes. As recently pointed out by the American Lung Association:

"the increase in ultraviolet D radiation resulting from even a moderate loss in the total ozone column can be expected to result in a significant increase in peak ground based ozone levels." The ALA continued, "these high peaks will occur earlier in the day and closer to the populous urban areas in comparison to current experience, resulting in a significant, though quantitatively unspecified, increase in the number of people exposed to these high peaks." [30]

Further, tropospheric ozone is a greenhouse gas. Ozone absorbs infrared radiation and increased ozone concentrations in the troposphere will contribute to climate modification.

Carbon monoxide

Exposure to carbon monoxide results almost entirely from motor vehicle emissions. (In some localized areas, wood stoves also significantly affect CO levels.) While there has been progress in reducing ambient CO levels across Europe, Japan and the United States, the problem is far from solved. For example, approximately 35 major metropolitan areas in the U.S. with a population approaching 30 million currently exceed the carbon monoxide air quality standard. In fact, EPA indicated in Congressional testimony (February 1987) that as many as 15 areas in the U.S. may have intermittent carbon monoxide (CO) problems that could prevent attainment for many years [26].

The CO problem is important because of the clear evidence relating CO exposure to adverse health effects. For example, in a recent assessment conducted under the auspices of the Health Effects Institute, it was concluded that:

"These findings demonstrate that low levels of COHb produce significant effects on cardiac function during exercise in subjects with coronary artery disease." [31]

Further, in another recent study of tunnel workers in New York City, the authors noted:

"Given the magnitude of the effect that we have observed for a very prevalent cause of death, exposure to vehicular exhaust, more specifically to CO, in combination with underlying heart disease or other cardiovascular risk factors could be responsible for a very large number of preventable deaths." [32]

In addition, as noted earlier, recent evidence indicates that CO may contribute to elevated levels of tropospheric ozone [18].

Oxides of nitrogen

NOx emissions from vehicles and other sources produce a variety of adverse health and environmental effects. NOx emissions also react chemically with other pollutants to form ozone and other highly toxic pollutants. Next to sulfur dioxide, NOx emissions are the most prominent pollutant contributing to acidic deposition.

Exposure to nitrogen dioxide (NO_2) emissions is linked with increased susceptibility to respiratory infection, increased airway resistance in asthmatics, and decreased pulmonary function [33]. While most areas of the U.S. currently attain the annual average national air quality standard, short term exposures to NO_2 have resulted in a wide ranging group of respiratory problems in school children (cough, runny nose and sore throat are among the most common) as well as increased sensitivity to bronchoconstrictors by asthmatics [34, 35].

The World Health Organization concluded that a maximum 1 hour exposure of 190 - 320 micrograms per cubic meter (0.10 - 0.17 ppm) should be consistent with the protection of public health and that this exposure should not be exceeded more than once per month. The State of California has also adopted a short term NO_2 standard, 0.25 ppm averaged over one hour, to protect public health.

Oxides of nitrogen have also been shown to affect vegetation adversely. Some scientists believe that NOx is a significant contributor to the dying forests throughout central Europe [36]. This adverse effect is even more pronounced when nitrogen dioxide and sulfur dioxide occur simultaneously. Further, nitrogen dioxide has been found to cause deleterious effects on a wide variety of materials (including textiles dyes and fabrics, plastics, and rubber) and is responsible for a portion of the brownish colorations in polluted air or smog.

Acid deposition results from the chemical transformation and transport of sulfur dioxide and nitrogen oxides. NOx emissions are responsible for approximately one third of the acidity of rainfall. Recent evidence indicates that the role of NOx may be of increasing significance with regard to this problem:

"Measurements of the nitrate to sulphate ratio in the atmospheric aerosol in southern England have shown a steady increase since 1954. The nitrate content of precipitation averaged over the entire European Air Chemistry Network has steadily increased over the period 1955 to 1979. The nitrate levels in ice cores from South Greenland have continued to increase steeply from 1975 to 1984, whilst sulphate has remained relatively constant since 1968. The 'Thousand Lake Survey' in Norway has recently revealed a doubling in the nitrate concentration of 305 lakes over the period 1974, 1975 to 1986, despite little change in pH and sulphate." [37]

Several acid deposition control plans have targeted reductions in NOx emissions in addition to substantial reductions in sulfur dioxide. Furthermore, the ten participating countries at the 1985 International Conference of Ministers on Acid Rain committed to "take measures to decrease effectively the total annual emissions of nitrogen oxides from stationary and mobile sources as soon as possible". [38]

Conclusions

Motor vehicle emissions of HC, CO and NOx, therefore, can be seen as a major source not only of climate modification but also of adverse health and other environmental effects from ground level pollution. In addition, tropospheric pollution and climate modification have been found to be directly linked by a variety of mechanisms. To deal with these problems in a coordinated fashion requires the minimization of carbon monoxide, carbon dioxide, hydrocarbons, nitrogen oxides and chlorofluorocarbons.

On a global scale, emissions of these pollutants depends on the number of vehicles in use and their emission rates. In turn, their actual emission rates depend on their fuel efficiency and their use of control technologies which are available.

Increased population and economic activity in the future holds the potential to increase the problem. Whereas the number of people in Europe and the U.S. is increasing slowly, the global population is expected to double (compared to 1960 levels) by the year 2000, driven by more than a doubling in Asia and an almost 150% increase in Latin America. Beyond the overall growth in population, an increasing portion of Asia's and South America's people are moving to cities, driving up the global urban population. One result is that global automobile production and use are projected to continue to grow substantially over the next several decades.

The Toronto conference on global warming which took place last year concluded that in the short term, *i.e.*, over the next 15 years, global CO_2 reductions in the order of 20% will be necessary to restrain global warming; over the longer term, reductions of approximately 50% *from current emissions rates* appear necessary to stabilize the global climate. Experience gained during the 1970s and 1980s in the U.S. suggests that the dual goals of low emissions (CO, HC and NOx) and improved energy efficiency (and therefore lower CO_2) are not only compatible but mutually reinforcing as illustrated in Fig. 2. However, continuing air pollution and the emergence of the global warming phenomena, indicate that it is now time to look for the next technological leap forward.

Fuel cells: the energy/environment solution [39 - 52]

Fuel cell technology is essentially just a battery, using an external supply of fluids as its energy source, and solids to separate those fluids, connected to an electric motor. Unlike a battery, however, a fuel cell does not run down or require recharging; it will operate as long as both fuel and oxidant (oxygen in air) are supplied to the electrodes and the electrodes remain separated by the electrolyte. The electrodes act as catalytic reaction sites where the fuel and oxidant are electrochemically transformed, producing d.c. power, water, and heat. Since fuel cells are not limited by Carnot's theory of heat engines (as are all conventional engines), their potential efficiency is much greater.

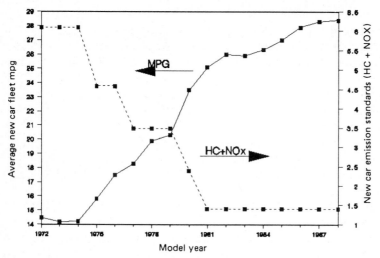

Fig. 2. U.S.A. emissions and fuel economy trends.

Studies have demonstrated that fuel cells have the potential to approximately double vehicle fuel efficiency. Less fuel consumption for the global fleet is the real key to lower CO_2 emissions. This energy conservation can take the form of less miles driven per vehicle per year, or less total vehicles, or more efficient vehicles; ultimately it will probably require some combination of each.

At the same time, fuel cells have the potential to substantially lower if not eliminate some of the conventional pollutants. For example, NOx production is due to interactions between the oxygen and nitrogen in the air at high temperatures; since fuel cells operate at much lower temperatures than conventional combustion engines, they should emit less NOx. Hydrocarbon production is mainly due to incomplete combustion; fuel cells do not rely on combustion, except in auxiliary systems which produce warm gases which are recycled into the system.

Potential impact of fuel cells on CO_2 emissions

Figure 3 shows the likely trend in fuel consumption by the worldwide vehicle fleet over the next forty years under alternative growth scenarios and based on current vehicle fuel economy trends. Based on historical trends, global VMT growth of at least 2% per year is likely. Should this occur, Fig. 4 shows the potential impact on global CO_2 emissions from the vehicle fleet.

Figure 5 illustrates the potential impacts of several alternative strategies to address this concern. It shows that a 1% annual improvement in new vehicle fuel efficiency starting in 1990 can start to *reduce the rate of growth in carbon dioxide emissions but is not sufficient to reverse the trend.* Alternatively, conversion of 1% of the new vehicle fleet by the year 2000 to

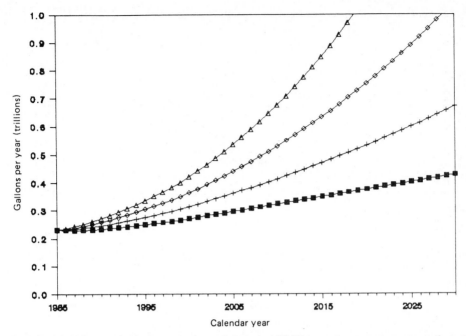

Fig. 3. Global auto fuel consumption, alternative VMT growth rates: ■, no growth; +, 1% growth; ◇, 2% growth; △, 3% growth.

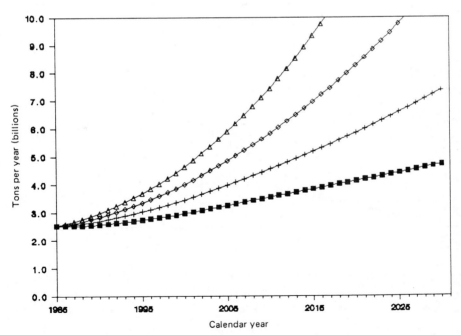

Fig. 4. Global auto CO_2 emissions, alternative VMT growth rates: ■, no growth; +, 1% growth; ◇, 2% growth, △, 3% growth.

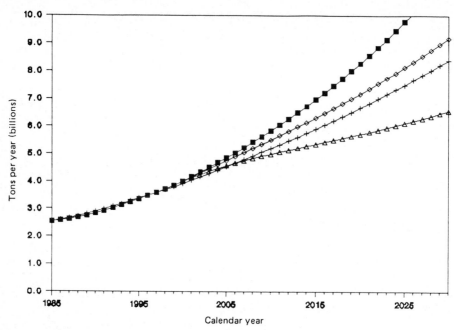

Fig. 5. Global auto CO_2 emissions, alternative strategies: ■, base; +, 1% efficiency gain; ◊, 1% fuel cells; △, 3% fuel cells.

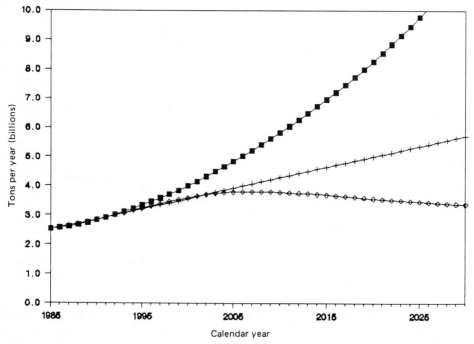

Fig. 6. Global auto CO_2 emissions, combined strategies: ■, base; +, 2% efficiency gain; ◊, 3% fuel cells.

operate with fuel cells, increasing at a rate of 1% per year after 2000, can also start to lower the rate of growth in CO_2 emissions. If this rate of conversion could be increased to 3% per year, the overall impact would be even more substantial. However, none of these strategies alone are sufficient to overcome expected vehicle growth.

Figure 6 shows the potential impact of a combination of strategies. Specifically, if 'conventional' new vehicles were to improve at a rate of 2% per year starting in 1990 and 3% of the fleet were to convert to fuel cells and achieve twice the efficiency of conventional vehicles starting in the year 2000, it would be possible to actually start to reduce the global CO_2 remissions from the global vehicle fleet.

The overall policy to address air pollution and global warming

The correct policy to minimize local and global environmental problems from vehicles in the future would have the following elements.

(1) Stringent emission standards for CO, HC and NOx such that all new vehicles sold around the world are equipped with 'state of the art' catalytic emissions controls. This state of the art should continue to move toward lower and lower levels. In the short term, at a minimum, these levels should be no higher than those already adopted by the State of California. Longer

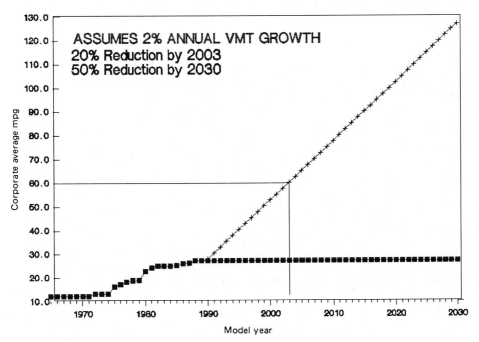

Fig. 7. CAFE requirements to achieve CO_2 target compared to 1988 base: ■, base m.p.g.; +, improved m.p.g.

term, even lower levels should be mandated as advanced technologies such as fuel cells enter production.

(2) Carbon dioxide standards which will be sufficient to lower global fleet emissions by 20% in the short term and 50% in the longer term (taking into account not only the direct emissions but also the emissions associated with the entire extraction, manufacture and distribution of the fuel). As shown in Fig. 7, for the U.S. this likely means fuel efficiency levels for gasoline fueled cars approaching 60 m.p.g. by the year 2000 and 125 m.p.g. by 2030.

Fuel cells clearly have great potential to play a significant role in addressing both these objectives and therefore should be receiving much greater attention. While energy concerns should continue to be a major motivation, local, regional and global air pollution concerns may be an even more significant reason to aggressively pursue this advancement.

References

1 Atmospheric changes called irrefutable; scientists urge greater research effort, *International Environment Reporter*, Aug. 10, 1988.
2 R. J. Ramanathan, *et al.* Trace gas trends and their potential role in climate change, *J. Geophys. Res.*, *90* (1985) 5547 - 5566.
3 J. Hansen and S. Lebedeff, Global surface air temperatures: update through 1987, *J. Geophys. Res.*, (1988) to be published.
4 The great flood of heat: 42 days and 42 nights, and life is altered, *N.Y. Times*, Aug. 14, 1988.
5 Atmospheric changes called irrefutable; scientists call for greater research effort, *Environment Reporter*, July 15, 1988.
6 R. A. Kerr, Is the greenhouse here?, *Science*, *239* (1988) 559 - 561.
7 High global temperatures indicate trend, not chance occurrence, NASA official says, *International Environment Reporter*, July 13, 1988.
8 Global warming, Letters, *Science*, 26 Aug., 1988.
9 Atmospheric ozone: assessment of our understanding of the processes controlling its present distribution and change, Geneva, *WMO Global Ozone Research and Monitoring Project, Report No. 16.*
10 F. S. Rowland and R. Watson, Committee on Environment and Public Works, U.S. Senate, Washington, DC, March 30, 1988 (Testimony).
11 A. Volz and D. Kely, Evaluation of the montsouris series of ozone measurements made in the nineteenth century, *Nature (London)*, *332* (1988) 240 - 243.
12 S. A. Penkett, Atmospheric chemistry: increased tropospheric ozone, *Nature (London)*, *332* (1988) 204.
13 Photochemical Oxidant Episodes, Acid Deposition and Global Atmospheric Change. The Relationships with Emission Changes of Nitrogen Oxides and Volatile Organic Compounds, Oystein Hov, Norwegian Institute for Air Research, February 1988.
14 Greenhouse Effect, Sea Level Rise and Coastal Wetlands, U.S. EPA, 1988.
15 D. R. Blake before the Committee on Energy and Natural Resources, U.S. Senate, Washington, DC, Nov. 9, 1987 (Testimony).
16 A. M. Thompson and R. J. Cicerone, Atmospheric CH_4, CO and OH from 1880 to 1985, *Nature (London)*, *321* (1988) 143 - 150.
17 V. Ramanathan, The greenhouse theory of climate change: a test by an inadvertent global experiment, *Science*, 15 April, 1988.

18 G. J. MacDonald, *The Greenhouse Effect and Climate Change*, presented to Env. & Public Works Committee, Jan. 28, 1987.

19 M. A. K. Khalil and R. A. Rasmussen, Carbon monoxide in the earth's atmosphere: indications of a global increase, *Nature (London), 332* (1988) 245.

20 M. A. DeLuchi *et al.*, Transportation fuels and the greenhouse effect, *Transportation Research Record*, submitted for publication.

21 *Regulatory Impact Analysis: Protection of Stratospheric Ozone*, Environmental Protection Agency, Washington, DC, Dec. 1987.

22 *The Transport Sector and Global Warming*, Background Study of OTA Report, Parsons, May 31, 1989.

23 *An Emission Inventory for SO₂, NOx and VOCs in North Western Europe*, Lubkert, de Tilly, Organization for Economic Cooperation and Development, 1987.

24 *OECD Environmental Data*, Organization for Economic Cooperation and Development, Paris, 1987.

25 G. J. McRae, Written statement prepared for U.S. House of Representatives Committee on Energy and Commerce, Subcommittee on Health and the Environment, Feb. 9, 1987.

26 L. M. Thomas, Testimony before the Subcommittee on Health and the Environment, Committee on Energy and Commerce, Washington, DC, Feb. 19, 1987.

27 U.S. Environmental Protection Agency, *National Air Quality and Emissions Trends Report, 1986*, Feb. 1988.

28 Tighter ozone standard urged by scientists, *Science*, 24 June, 1988.

29 J. J. Mackensie and M. El Ashry, *Ill Winds Pollution's Toll on Trees and Crops*, World Resources Institute, Sept. 1988.

30 American Lung Association, Comments to EPA, 1988.

31 Health Effects Institute, 1988.

32 F. D. Stern *et al.*, Heart Disease Mortality Among Bridge and Tunnel Officers Exposed to Vehicular Exhaust, NIOSH.

33 Lindvall, *Health effects of nitrogen dioxide and oxidants*, Department of Environmental Hygiene, National Institute of Environmental Medicine and Karolinska Institute, March 17, 1982.

34 Orehek, *et al.*, Effect of short term, low level nitrogen dioxide exposure on bronchial sensitivity of asthmatic patients, *J. Clin. Investig., 57* (Feb. 1976).

35 Mostardi *et al.*, The University of Akron study on air pollution and human health effects, *Archives Environ. Health*, (Sept./Oct. 1981).

36 Whetstone and Rosencranz, *Acid Rain in Europe and North America*, Environmental Law Institute, 1983.

37 A. better way to control pollution, Derwent, *Nature (London), 331* (18 Feb., 1988).

38 *Int. Conf. Ministers on Acid Rain, Ottawa, Canada, March 1985.*

39 T. C. Benjamin and E. H. Camara, *The Fuel Cell: Key to Practically Unlimited Energy*, Foote Mineral Company, Exton, PA, 1985.

40 P. J. Brown, K. F. Barber and R. Kirk, *SAE J. Electric Vehicles' Transportation Potentials, 98*, (8).

41 Platinum in fuel cell development, *Platinum Met. Rev.*, (1989).

42 Los Alamos National Laboratory, Engineering Div., *Fuel Cells for Extraterristrial and Terrestrial Applications, 136* (2).

43 *J. Electrochem., Soc.*, The Electrochemical Society, Inc., Feb. 1989.

44 P. J. Werbos, Oil dependency and the potential for fuel cell vehicles, *SAE Technical Paper Series*, May 18 - 21, 1987.

45 S. S. Penner, *Assessment of Research Needs for Advanced Fuel Cells*, U.S. Department of Energy, Springfield, VA, Nov. 1985.

46 M. Krumpelt and R. Kumar, *An Assessment of Fuel Cells for Transportation Applications*.

47 R. J. Kevala and D. M. Marinetti, Fuel cell power plants for public transport vehicles, *SAE Technical Papers*.

48 C. V. Chi, D. R. Clenn and S. C. Abens, Air Cooled Phosphoric Acid Fuel Cell/Ni Cd Battery Powered Bus, August 1989.

49 P. C. Patil and J. R. Huff, Fuel Cell/Battery Hybrid Power Source for Vehicles.

50 P. Patil, C. Christianson and S. Romano, Integration of a fuel cell/battery power source in a small transit bus system, *IECEC*, *88* (Apr. 29, 1988).

51 R. L. Rentz, C. L. Hagey and R. S. Kirk, Fuel cells as a long range highway vehicle propulsion alternative, *IECEC*, (Aug. 1986).

52 H. S. Murray, Fuel cells for bus power, *SAE Technical Paper Series*, Nov. 1986.

Journal of Power Sources, 29 (1990) 29 - 35

FUEL CELL DEVELOPMENTS IN JAPAN

NOBORU ITOH*

Faculty of Science and Engineering, Chuo University, 1-13-27 Kasugacho, Bunkyoku, Tokyo 112 (Japan)

Overview

Research to develop fuel cells in Japan was initiated by research institutions beginning about 1955 and is therefore backed by roughly 30 years of experience. In the years after 1965, the principal result of research was the commercialization of dissolved fuel type cells using methanol and other substances as the fuel for utilization for use as a compact power source in energizing radio relay stations and other systems. However, it was only during a short span of less than a decade that particularly significant progress was achieved in related research.

Central to research on these fuel cells is the Moonlight Project implemented in 1981 by the Agency of Industrial Science and Technology (AIST) of the Ministry of International Trade and Industry (MITI). At about the same time, research conducted broadly by governmental and private sector research institutions, including the introduction of American made experiment plants, served as the foundation, together with the Moonlight Project, for promoting fuel cell research to today's high level. Intensive research is being advanced in connection with three types of fuel cells: phosphoric acid fuel cell (PAFC), molten carbonate fuel cell (MCFC) and solid oxide fuel cell (SOFC).

Government Research and Development (R&D) — Moonlight Project

The development of fuel cells in Japan is proceeding smoothly. Research on the phosphoric acid type is presently being advanced with a demonstration plant, and research on the carbonate type is approaching the stage of a demonstration system. These development activities are also being advanced by the private sector, but the government's R&D activities to develop related domestic technologies are playing the central role.

AIST started research on fuel cells in fiscal year 1981 as a link to its "Large-scale R&D Projects for Energy Conservation" (Moonlight Project), and is presently engaged in research mainly on the molten carbonate fuel cell, but also on the phosphoric acid and solid oxide fuel cells. Figure 1

---

*Formerly Director General of Energy Conversion and Storage Dept. of NEDO.

0378-7753/90/$3.50

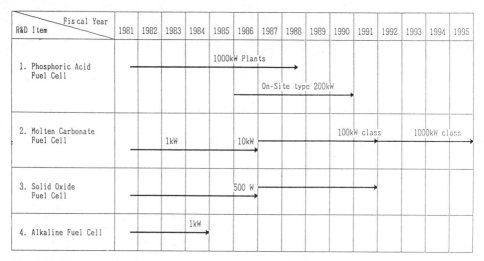

Fig. 1. R&D timetable of fuel cell power generation technology.

shows the R&D program for developing fuel cell power generation technologies.

The fuel cell R&D project had originally been structured on a 10-year program. In March, 1987 the basic plan was changed to a 15-year program up to fiscal year 1995 owing to new development research on the molten carbonate fuel cell. Subsequently, the total R&D budget was increased to about ¥57 billion. In order to advance these R&D activities most effectively, a Fuel Cell Power Generation Subcommittee has been established in the Industrial Technology Deliberation Council, an advisory organ to the Minister of International Trade and Industry. This subcommittee will deliberate on vital matters such as program evaluation and research result evaluation.

Outline of Government R&D activities

A project to develop a phosphoric acid type 1000 kW class power generation system is being advanced with the aim of developing fuel cell power systems for electric utilities. Two systems have been fabricated and installed — one for dispersed power generation (at Kansai Electric Power Co.'s Sakaiko Power Station) and the other for use as a substitute thermal power plant (at Chubu Electric Power Co.'s Chita 2nd Thermal Power Plant). The former was successfully put to 1000 kW power generation tests in September 1987, and the latter was put to the same tests in December 1987. Since then operation experiences were conducted and successfully terminated for both systems by the end of October and September 1988 respectively. As far as the generating power time was concerned, the former was over 2000 h, the latter over 1000 h. The achievement made by both plants during the operation research is shown in Table 1.

TABLE 1

Operational performance of 1000 kW pilot plant

Item	Dispersed generation use	Centralized generation use
Total power generation time max. cont. gen. time: 4 stacks 2 stacks	2045 h [118 h (500 ~ 750 kW)] [705 h (250 ~ 500 kW)]	1018 h [440 h (300 kW)]
No. cells per stack	515	485
Optimum temperature and pressure	190 °C 4 atm.	205 °C 6 atm.
Generated power output	697151 kW h	367583 kW h
No. generation	40	46
Reformer temperature rise time	3614 h	2410 h
Generating efficiency (1000 kW, HHV)	37.1%	38.4%
Start up time (hot start) (Catalyst temperature at start up)	5 h 1 min (250 °C)	4 h 20 min (359 °C)
Load following characteristics (load band)	±11%/m (125 ~ 450 kW)	±10%/m (150 ~ 400 kW)
Stopping time: ordinary emergency	47 min 10 s	60 min 10 s
NOx (at 1000 kW)	10 ppm	8.5 ppm
Noise (A level: at 1000 kW)	58 ~ 71 dB	59 ~ 74.5 dB

A project to develop a phosphoric acid type 200 kW class power generation system has been in progress since fiscal year 1986. This 5-year program is aimed at developing an on-site type fuel cell which would permit power systems to be installed flexibly in various kinds of power demand regions. Specifically, two systems are under development — one using methanol as the fuel and designed to provide power systems for remote isolated islands, and the other using city gas as the fuel and designed to provide power systems for cogeneration. The former system will lend itself to a parallel operation with a diesel power generator using closely-matched load following characteristic methanol reformer, while the latter system will feature concurrent electricity heat supply, compactness and high reliability, for use by commercial establishments such as hotels and restaurants.

Meanwhile, with regard to the molten carbonate fuel cell, the New Energy Development Organization (NEDO) in 1986 succeeded in generating an output of 10 kW with a matrix type electrolyte (developer Hitachi Ltd.) and with a paste type electrolyte (developer Toshiba Corp.). The successful development of a 10 kW fuel cell offered some promising prospects for the enlargement and performance improvement of these fuel cells with existing materials and manufacturing technologies, raising the operating pressures, improving gas utilization efficiencies and enabling large-scale stacks. The second-phase program for developing the molten carbonate fuel cell is to start at the beginning of fiscal year 1987 in the Moonlight Project. The plan is to strive to resolve these previously mentioned technological problems while developing a 100 kW fuel cell stack in the first half 5-year period up to fiscal 1991. This will be followed by the development of technologies for fabricating a power generation plant system on the basis of these results, and in the latter 4-year period up to fiscal 1995, to develop a 1000 kW class pilot plant.

In January 1988, the Technology Research Association for MCFC Power Generating System was established for the purpose of developing the system technology of a 1000 kW class pilot plant including peripheral equipment, balance of plant and other related technology.

The internal reforming system capitalizes on the high temperature that is a distinct characteristic of the molten carbonate fuel cell, which enables fuel gas for utilization in power generation to be generated through catalytic reforming of natural gas inside the cell, making the use of a separate reformer unnecessary. The development of this internal reforming system still lies in the laboratory experimental stage compared with external reforming systems, however NEDO envisages the IR-MCFC development of a few 10 kW scale in the second-phase program.

The Government Industrial Research Institute in Osaka is conducting research in search of new fuel cell materials and developing technologies for evaluating the properties of various kinds of materials.

With regard to the solid oxide fuel cell, the Electrotechnical Laboratory is engaged in research to develop a 500 W scale cell based on the tubular concept, while repeating experiments to increase the cell's durability by improving the cell's structural soundness of components, and upgrading the cell's performance by developing high quality elementary materials. Meanwhile, the National Chemical Laboratory for Industry is conducting research to develop the planar type in the same project. More specifically, research is being advanced to develop a composite three-layered film combining anode, cathode and electrolyte in a compact assembly. NEDO started the feasibility study on SOFC development in a three-year program for the purposes of manufacturing high power density cell/stacks, compact SOFC concept and long-life performance. Figure 2 shows the R&D setup in the Moonlight Project.

Private sector R&D activities

Research on PAFC in Japan is being advanced by the private sector with electric power companies and gas companies assuming the central roles. Tokyo Gas Co., Ltd and Osaka Gas Co., Ltd. participated in America's GRI

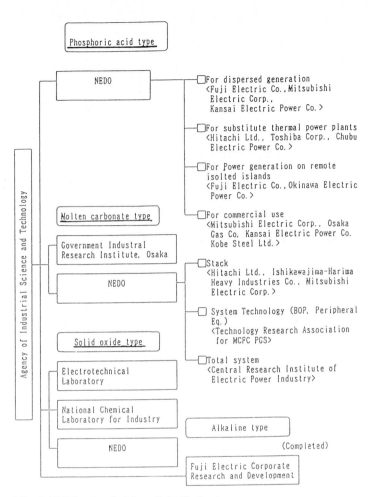

Fig. 2. R&D setup in Moonlight Project.

Project, and were engaged in operational research on fuel cells as the field test program. Osaka Gas, in particular, has completed the operational test in a family restaurant; the plant had run for 15 600 h. Meanwhile, Tokyo Electric Power Co., Inc. has conducted running tests since 1983 on a 4.5 MW fuel cell system developed by UTC (U.S.A.) and achieved the system's rated output.

R&D activities are being advanced intensively by various corporations to develop domestic fuel cells; for example, Tohoku Electric Power Co. is using 50 kW PAFC developed by Fuji Electric Co. and Hokkaido Electric Power Co. is using methanol fueled 100 kW PAFC by Mitsubishi Electric Co., and by introducing fuel cells developed in the United States; one typical example is Tokyo Electric Power Co. with a IFC 200 kW on-site fuel cell. Table 2 shows the outline of private utilities fuel cell power plant activities.

TABLE 2

Outline of private utilities activity

Name of company	Outline of plant	Manufacturer	Remarks
Tokyo Electric Power Co.	220 kW, NG, air cooling	Sanyo Elec.	Aug.'86 construction
	200 kW, NG, water cooling		Sept.'87 operation
		IFC	Oct.'88 operation study
	200 kW, NG, water cooling	IFC	Feb.'89 heat suppling test
	11 MW, NG, water cooling	Toshiba/IFC	Oct.'89 fabrication
			Apr.'89 installation
			Dec.'90 operation start
			Jan.'91 demo operation
Tohoku Elec. Power Co.	50 kW, NG/LPG, 190 °C, boiling water cooling	Fuji Elec.	Oct.'85 construction
			Mar.'87 100% operation
			Mar.'90 be terminated
Hokkaido Elec. Power Co.	100 kW, methanol, 190 °C, boiling water cooling	Mitsubishi Electric	Nov.'87 - Mar.'89 operation
			Accumlated OP. hours 4575 h 330 MW h, co-generation use
Tokyo Gas Co.	50 kW, NG, 190 °C, 1 atg, boiling water cooling	Fuji Elec.	Very compact design Grid connected
	100 kW, NG, 190 °C, 1 atg, boiling water cooling	Hitachi Ltd.	Same as above
Osaka Gas Co.	200 kW, NG, 190 °C, 1 atg, boiling water cooling	IFC	Mar.'89 installation
			Apr.'89 operation
			Mar.'91 be terminated

On the other hand, the molten carbonate fuel cell technology is advancing with stack manufacturers presently engaged in research to develop their own unique 10 kW fuel cell systems in the same manner as the Moonlight Project.

During the period from the end of March through April of 1987, three companies Ishikawajima-Harima Heavy Industry, Mitsubishi Electric and Fuji Electric Co. R&D, which were engaged in the development of related elementary technologies, also succeeded in developing a unique 10 kW fuel cell stack. This feat advanced domestic research on the molten carbonate fuel cell from the stage of laboratory experiments to the stage of full scale development of commercial type fuel cell. This brought Japan up to approximately the same technological level with the United States which, up to this point, had a good head start.

The successful development of the 10 kW fuel cell offered some attractive prospects for the enlargement and performance improvement of these fuel cells with existing materials and manufacturing technologies. However, various problems still remain to be resolved, including the development of technologies for extending service life expectancies, raising the operation pressures, improving gas utilization efficiencies and enabling large scale stacks.

The second-phase program for developing the molten carbonate fuel cell is to be started beginning fiscal year 1987 in the Moonlight Project.

In solid oxide fuel cell research and development, some private concerns are also very active. Mitsubishi Heavy Industries, for example, has developed a fuel cell capable of generating an output of about 10 W with a single cell unit, and is scheduled to conduct evaluation experiments over a period of one year jointly with the Tokyo Electric Power Co. The company has also drafted a plan to venture into research to develop kilowatt class unit cells within the next few years.

Incidentally, Tokyo Gas and Osaka Gas concluded a contract in 1986 with Westinghouse for purchasing stacks, and they have conducted demonstration tests on a 3 kW stack since the end of 1987.

Recently some developers such as Fuji Electric Co., Fujikura and Sanyo have begun elemental research work for SOFC.

Conclusions

It is quite meaningful that the domestic technology on the phosphoric acid fuel cell power generation system has been rapidly upgraded from the 10 kW scale to the 1000 kW scale in a short period of time, getting closer to the highest technological level in the world, obtaining much necessary information in the development of commercial applications and clarifying the position as the fourth power generation method.

A lot of information obtained from the national R&D Project is considered to be utilized not only in the development of PAFC but also in the development of other fuel cells.

Hereafter, the second generation and third generation fuel cells should be encouraged more in the R&D Program for future high power and high efficiency generation methods.

The prospect for steady development of the phosphoric acid fuel cell power generation system for utility use and commercial application has become clear from the national R&D Project. Also, overseas nations have great expectations of Japan's move toward the development of fuel cell commercial applications. Therefore, it is necessary to clarify the next project plan — multi-MW demonstration plant — soon and accelerate activities toward the development of commercial applications under much closer cooperation among industry, academia and government than ever before.

Journal of Power Sources, 29 (1990) 37 - 45

FUEL CELL APPLICATIONS AND MARKET OPPORTUNITIES

KAREN TRIMBLE and RICHARD WOODS

Gas Research Institute, 8600 West Bryn Mawr Avenue, Chicago, IL 60631 (U.S.A.)

Fuel cells represent an exciting generation technology for the next decade. Their efficiency, modularity, environmental characteristics and siting flexibility will permit their use in a variety of applications worldwide. Market opportunities for fuel cells will depend, to a large extent, on their individual operating characteristics. With phosphoric acid fuel cell technology positioned for commercial introduction in the early 1990s, the Gas Research Institute is currently focused on the development of advanced fuel cell technologies for expanded cogeneration applications in the commercial and light industrial market sector.

Several of the more important benefits of advanced fuel cell technology for cogeneration applications are high conversion efficiencies (>45% HHV for practical systems), high quality by-product heat, and the potential for direct natural gas utilization. The high efficiencies of advanced fuel cells are particularly attractive where thermal to electric ratios are low. In these applications, the higher electrical efficiencies permit fuel cell systems to be sized larger than competing cogeneration systems in a given application, hence, servicing a greater kilowatt load with gas. Additionally, the high exhaust temperatures enhance coupling with thermally driven chillers, again displacing electric peaking demand.

Table 1 identifies a number of potential markets for fuel cell generators. The residential, aerospace/military and transportation markets/applications are beyond the scope of this paper.

Figure 1 attempts to summarize the applications of the fuel cell types and their support within the National Fuel Cell Coordinating Group.

All told, there are about 3300 individual utilities in the first four categories in Table 1. An additional 2800 cogenerators/independent power

TABLE 1

Fuel cell market sectors

Electric utility power generation
Independent power production
Industrial cogeneration
On-site/commercial cogeneration
Aerospace/military
Residential cogeneration
Transportation

NATIONAL FUEL CELL COORDINATING GROUP	FUEL CELL TECHNOLOGY				
	AFC	SPFC	PAFC	MCFC	SOFC
Department of Energy (DOE)		Transportation	Electric Utility Cogeneration - Commercial Transportation	Electric Utility Cogeneration - Commercial - Industrial	Electric Utility Cogeneration - Commercial - Industrial
National Aeronautics and Space Administration (NASA)	Space Power	Space Power			
Department of Defense (DOD)	Military				Military
Electric Power Research Institute (EPRI)			Electric Utility IPP	Electric Utility Cogeneration IPP	Electric Utility Cogeneration IPP
Gas Research Institute (GRI)			Cogeneration - Commercial	Cogeneration - Commercial - Industrial	Cogeneration

Fig. 1. Application interest by fuel cell type for national funding organizations.

producers also supply electricity, though generally to a single customer (themselves). Though cogenerators/independent power producers still serve relatively few customers, they have been a most important category in terms of new capacity additions in recent years, accounting for 14 000 MW of the total capacity added since 1980.

The commercial and industrial sector in the United States represents a large potential market for fuel cells. The commercial sector includes four million sites representing approximately 175 000 MWe of cumulative peak demand. There are many applications in the commercial sector which have sufficient electric and thermal loads to make fuel cell cogeneration economically attractive. Table 2 presents a number of these applications and some general characteristics of each. This market is projected to grow by 2 - 4% per year with more than 50% of the market being comprised of office buildings, stores and appartments.

An economic screening of the potential commercial market reveals that only a portion is economically attractive or feasible. Based on inputs regarding local energy rates, operating modes, and sizing strategies, the economically feasible market size has been estimated. If the long range objective of a full catalog of grid-independent power plants at installed costs of $1000/kW is realized, an economically feasible market of 18 000 MW or about 10% of the commercial sector's population results. If 100% of this economically feasible market were captured, it would represent 0.52 trillion cubic feet (U.S.) of additional annual gas sales.

Many U.S. cities examined exhibit commercial retail energy prices that make gas-fueled cogeneration economically feasible for systems with

TABLE 2

Commercial applications

Application	No. sites	Average electrical use (kW)	Average thermal load (MMBtu)
Hospitals	7100	300 - 800	0.8 - 2.0
Hotels/motels	56470	100 - 850	0.3 - 3.0
Supermarkets	29550	240 - 300	0.2 - 0.3
Restaurants (18 - 24 h)	20500	60 - 90	0.4 - 0.6
Nursing homes	14950	60 - 70	0.4 - 0.6
Large office buildings	24000	200 - 800	1.1 - 3.5
Apartment complexes (>20 units)	54000	50 - 400	0.3 - 2.5
Shopping centres	7820	210 - 2200	0.4 - 3.5
Laundries	75000	20	0.6
Educational facilities	13000	500 - 2000	0.5 - 1.0

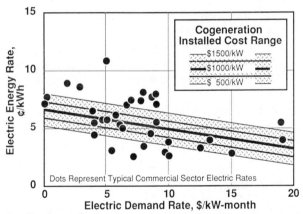

Base Line Assumptions:
30% Electric Efficiency; 75% Overall Efficiency; 80% Boiler Efficiency; 20% Annual Capital Charge Rate; 80% Capacity Factor; 50% Heat Utilization; 0.5 ¢/kWh O&M Rate; & $4.00/MMBtu Gas Price.

Fig. 2. Self-generation economics; commercial sector or on-site cogeneration.

installed costs of approximately $1000/kW. Figure 2 illustrates the commercial viability of these systems at a number of typical cities. Efficiencies in the range of 30% (HHV) are typical of reciprocating engines and small turbines in this size range. Experience to date indicates that installed costs for these competing systems in the range of $1000 - 1500/kW are typical for small cogeneration systems in the 50 - 500 kW range. These costs reflect the high costs for engineering, site assembly, and installation.

GRI has projected that on-site fuel cells, under mature market conditions, will have an installed cost of approximately $1000/kW. This is based on economies of scale for a continuous production line operation supplying several hundreds of megawatts per year. During the production of the introductory units, the installed costs are projected to be relatively expensive ($2000 - 3000/kW).

To realize the immense potential that exists within this sector, technology and market developments are required to reduce first cost and increase system utilization efficiencies. Cost effective integration of absorption chillers is the key to effective utilization of by-product heat in many commercial applications and thus can enhance market penetration. Other uses of by-product heat include water heating, space heating and steam production. Fuel cell systems for commercial sector applications are expected to range from 25 to 1000 kW in size. Overall efficiencies approaching 80% have been projected for these systems.

The industrial market is approximately equal in size in terms of peak electrical demand to the commercial sector. This market segment includes about 300 000 sites with 20 000 of those accounting for over 90% of the energy use. Fifty percent of this market is represented by the chemicals and primary metals industries. The small industrial market (<2 MWe) represents about 10% of the industrial market or approximately 15 000 MWe peak demand — the food industry comprises about 20% of this market. Advanced fuel cells are expected to be used in municipal waste treatment plants, breweries, chemical plants, paper making, petroleum and metals refining and chlor-alkali production. PAFC technology is limited in this application due to the lower quality heat available and strong competition from conventional technologies.

The industrial sector is characterized as having relatively high thermal to electrical needs which minimize the economic advantages of high electric generation technologies. Figure 3 illustrates the impact of heat utilization

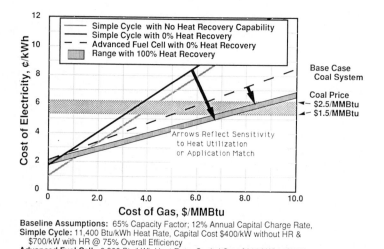

Baseline Assumptions: 65% Capacity Factor; 12% Annual Capital Charge Rate,
Simple Cycle: 11,400 Btu/kWh Heat Rate, Capital Cost $400/kW without HR &
 $700/kW with HR @ 75% Overall Efficiency
Advanced Fuel Cell: 6,300 Btu/kWh Heat Rate, Capital Cost $900/kW and 75%
 Overall Efficiency

Fig. 3. Impact of heat recovery; industrial cogeneration.

for both a simple cycle turbine and an advanced fuel cell. As can be seen, the advanced fuel cell is much less sensitive to heat utilization.

Characteristics other than efficiency and cleanliness which enhance industrial fuel cell use are: (1) a fuel cell power plant can provide d.c. power at lower cost than a.c. power, (2) it may be possible to utilize industrial process off-gases in the fuel cell rather than combust the gases as is usually done, and (3) the fuel cell process might be directly integrated into industrial processes, particularly in the chemical process industry.

Figure 4 shows constant cost of electricity lines for various capacity factors for a typical natural gas fired combined cycle power plant. Plants such as this can be considered the main competition for advanced fuel cell systems in a similar size range. In order to be economical at all operating philosophies, the fuel cell installed cost must be below $800/kW at a heat rate of 8000 Btu/kWh or less. If capital costs exceed this target, fuel cells will only be competitive in more traditional baseload operation.

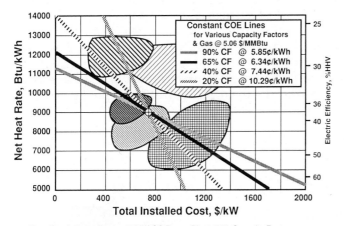

Baseline Assumptions: 14 MW CC Power Plant, 65% Capacity Factor; 12% Annual Capital Charge Rate, $750/kW Capital Cost, 9000 Btu/kWh Heat Rate, 0.026 ¢/kWh Variable and 10 $/kW-yr Fixed O&M, and $5.06/MMBtu Gas Price

Fig. 4. Gas-fueled economics; 14 MW combined-cycle.

We can surmise, with reasonable certainty, that fuel cell market introduction will occur with on-site units which are fueled by natural gas. This is due to many factors. First, and most obviously, these units are smaller, representing a required step in the development of the later, coal-fired systems. Second, the systems are simpler, not requiring pressurization and not requiring significant development of other system components. The commercialization of gas-fired systems, therefore, should be of low relative risk and more rapid than the federally funded coal-based systems primarily supported by the U.S. DOE.

With demand for electricity growing by about 2 - 3% per year, a significant burden will be placed on the demand for raw energy sources (natural gas, coal, nuclear and hydropower). Given the long lead times currently associated with construction of new generating capacity, it appears that currently planned additions will not be sufficient to meet end-use electricity demand in the mid to late 1990s. These lead times, tightening environmental constraints and the inability to site nuclear plants may force utilities to consider incremental capacity additions, initially with gas-fired combined-cycle plants, and later with more efficient, fuel cells coupled with coal gasifiers. Both molten carbonate and solid oxide technologies project market penetration in the 1995 - 2000 period, precisely at the time the need for new capacity rapidly increases.

Independent power production also has the potential to be an important source of power. The May 1989 Study by The American Public Power Association (APPA) includes analyses of the deregulation underway in the industry and concludes that fuel cells are becoming increasingly attractive to public power because of these changes. The competition to fuel cells in public power is principally purchased power and diesel generators because combined cycle or coal plants are too large for most public utilities, and joint action agencies lack control of sufficient transmission capacity to take advantage of scale of large plants. Tightening emissions regulations have made it increasingly difficult to site diesel generators, hence fuel cells become an important alternative to purchasing power in an unregulated environment. In the near term, therefore, public power may be the only significant electric utility market in the U.S. A profile of public power is provided in Fig. 5.

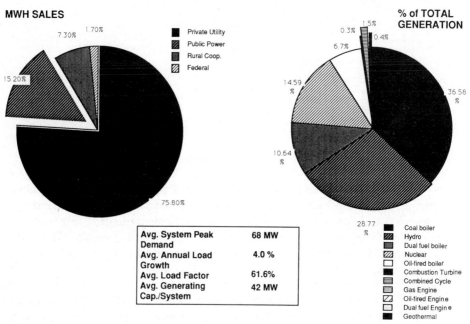

Fig. 5. Profile of public power.

Renegotiated power contracts plus load growth projections show a potential for 93 gigawatts over the 1996 - 2010 time frame. A 'likely early market' analysis was performed which results in a total of 14 GW over the time frame, or about 900 MW/year. Their study indicates that fuel cells can begin to enter this market at an allowable installed cost of $1500/kW. True competitive prices, however, are closer to $800/kW for PAFC or SOFC plants and $1000/kW for MCFC plants.

Independent power plants will be installed close to the load, increasing the probability for waste heat utilization. These plants will be approximately 50 MW or less in size. Dispersed generation has the potential for reducing capital costs and lead times required, and could reduce or postpone the need for new large-scale electric generating capacity. Advanced fuel cells, initially operating on natural gas, could fill this role. Initially the fuel cells will operate as load followers (1500 to 7000 h per year) but as the gap between coal and oil/gas prices widens, and as the need for baseload additions grows, fuel cell modules would be installed at central sites. Ganging several of these 20 - 50 MW generators and coupling them with a coal gasifier will complete the development and evolution of fuel cell power plants for gas and electric utility applications.

As discussed earlier, demand for electricity will outstrip supply leading to capacity shortfalls by the mid-1990s. Though present reserve margins are currently excessively high, planned capacity addition, when combined with retirements, will erode this reserve quickly. In 1995 the gap between total additions needed and total supply could be greater than 30 000 MW. By the year 2000, this figure could exceed 100 000 MW. Assuming that new units must be ordered five years in advance, the market for new units should take off in the mid to late 1990s.

An April 1987 study sponsored by EPRI indicates a 187 000 MW shortfall in electric capacity by the year 2010. In this study fuel cells were compared with competing technologies such as coal plants, combined-cycle plants and combustion turbines under a range of conditions for the capture of this market. For this comparison, fuel cells are estimated to cost $920/kW and have a heat rate of 7800 Btu/kWh. Under these base case assumptions, fuel cells capture 15 000 MW by the year 2010 or 8% of the market. With mature costs and heat rates meeting and/or exceeding this criteria, both carbonate and solid oxide fuel cell technologies project healthy futures if commercialization schedules can be adhered to.

The long-range market for fuel cells in the electric utility sector is based on integration with coal gasifiers and bottoming cycles for base load central station application. If fuel cell cost and performance targets are met, these will be the coal-fired option of choice as they are the most efficient and most environmentally of any coal consuming generation technology.

Figures 6 and 7 show economics of large scale utility power plants for natural gas and coal, respectively. Figure 6 suggests that it becomes difficult for advanced fuel cell power plants integrated with a coal gasifier to compete with natural gas fueled combined-cycle plants in the multi-hundred MW size

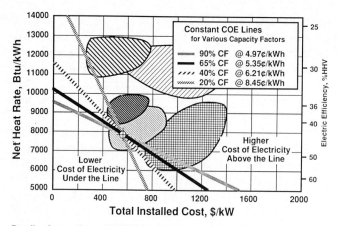

Baseline Assumptions: 380MW CC Power Plant, 65% Capacity Factor; 12% Annual Capital Charge Rate, $570/kW Capital Cost, 7800 Btu/kWh Heat Rate, 0.026 ¢/kWh Variable and 10 $/kW-yr Fixed O&M, and $5.06/MMBtu Gas Price

Fig. 6. Gas-fueled economics; 380 MW combined-cycle.

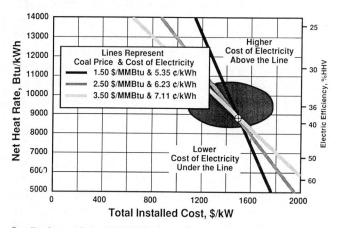

Baseline Assumptions: 380MW IGCC Power Plant, 65% Capacity Factor; 12% Annual Capital Charge Rate, $1500/kW Capital Cost, 8800 Btu/kWh Heat Rate, 0.17 ¢/kWh Variable and 40 $/kW-yr Fixed O&M

Fig. 7. Economics of coal-fueled power plants.

range. Their costs are approximately equal to combined-cycle systems for intermediate and baseload operation. Either environmental aspects or technical risk can tip the decision-making scales either way in the future. Allowable installed costs increase however, as shown in Fig. 7 for integrated gasifier combined-cycle power plants, making coal fired fuel cell plants more attractive when compared with like fuels.

Conclusions

The commercialization of fuel cells appears imminent with the debut of International Fuel Cell's PC-25 in 1991 - 1992. A number of orders have been placed for this unit, with expanding interest in Japan and regionally in the United States. Due to environmental constraints, the California market appears to be poised to take off. The South Coast Air Quality Management District has reviewed this technology and will host one of the first PAFC units sited. They estimate the California market at approximately 1000 units per year. Once doors such as these are opened, a window of opportunity will exist for improved and/or higher temperature fuel cells to penetrate the market.

Internal assessments performed at GRI generally show favorable economics for fuel cell systems if installed costs can reach their target goals. These observations are summarized below:

- In the commercial sector, cogeneration is competitive with electric energy changes in many cities. In this market, fuel cells can provide a one to two cent advantage over reciprocating engine based systems, mainly due to their higher electrical efficiency.

- In larger sizes and using higher efficiency gas fired equipment, gas fueled cogeneration is competitive with coal fueled power generation up to a natural gas price of $6 - 7/MMBtu. Advanced fuel cells, due to their high electrical efficiency, are much less sensitive to heat utilization making them easier to site. The opportunity also exists to integrate these systems directly into the process, especially in the chemicals industry and to provide direct current electricity for industries such as metals refining.

- The advanced fuel cells compete well with gas and coal fired combined cycle equipment in the smaller size ranges. For example, a 22.5 MW molten carbonate system can produce electricity at approximately the same price as a 100 MW combined cycle plant. When comparing equal size plants, advanced fuel cell power plants are superior to both combined cycle and steam injected gas turbines for both baseload and intermediate duty. Supplemental firing of these systems can provide added thermal to electric flexibility.

- For large scale utility plants, fuel cell power plants have a competitive cost of electricity when compared to alternate coal fueled technologies. A cost *versus* heat rate trade-off analysis shows that a 1600 Btu/kW h heat rate improvement approximately balances a $100/kW increase in capital costs at a fixed coal price of $1.5/MMBtu.

Fuel cell systems are thus expected to enter the market following successful demonstrations of commercial scale technology. These systems must, however, provide evidence of economic competitiveness and efficiency advantages when compared to conventional forms of energy service (purchased electricity and gas) and/or energy conversion technologies (*i.e.* engines, turbines, combined-cycle, etc.). It is therefore crucial for manufacturers and sponsors alike to concentrate efforts toward the development of low cost production techniques. This factor, above all others, will likely decide the future acceptance of fuel cell technology in the 1990s and beyond.

Journal of Power Sources, 29 (1990) 47 - 57

ELECTRIC POWER RESEARCH INSTITUTE'S ROLE IN DEVELOPING FUEL CELL SYSTEMS

DANIEL M. RASTLER

Generation and Storage Division, Electric Power Research Institute, 3412 Hillview Avenue, P.O. Box 10412, Palo Alto, CA 94303 (U.S.A.)

Introduction

The Electric Power Research Institute (EPRI) manages a collaborative research and development (R&D) program on behalf of the U.S. electric utility industry and its customers. Founded in 1972, the Institute is the nation's oldest major research consortium. Funded through voluntary contributions by over 600 member utilities, EPRI's technical staff directs research in a broad range of technologies related to the generation, delivery, storage and use of electricity. EPRI's staff provides member utilities with objective and current technical information on products and services ranging from complex systems such as coal gasification technology, state of the art electric vehicles, to software computer codes for improving plant productivity and efficiency. Special attention is given to developing products which are cost-effective and environmentally acceptable. Research and development expenditures in 1989 will be approximately $280 million (Fig. 1).

Since its inception, EPRI's role in developing new technologies for its customers has changed dramatically. Fifteen years ago, it was assumed that newly developed technologies would be wholeheartedly implemented by the utility industry and quickly adopted into their systems. A substantial effort has been made in developing and demonstrating new and improved power generation technologies, including the fuel cell, under the assumption that commercialization would happen without the need for EPRI support. However, during the past seven years, the electric utility business in the U.S. has changed significantly and this has impacted commercialization of new technology. Reduced load growth, an unpredictable fuel supply, high interest rates, and a difficult regulatory climate have contributed to the reluctance of utilities to pioneer new technology, especially if that requires accepting extraordinary costs and risk. Equipment suppliers, for similar reasons, have not been bullish on the electric utility industry as a business opportunity. Yet by the mid to late 1990s, the utility industry will need new technology options to meet the demand for electricity in a cost-effective, reliable and environmentally acceptable manner. To bring these new technologies to commercial fruition, EPRI's role has gone beyond focusing solely on technical and developmental issues; a better understanding of the industries' market-driven needs and customer requirements is now essential.

0378-7753/90/$3.50

Fig. 1. Research funded by EPRI. Total 1989 research and development budget: $280 million.

EPRI's role in developing fuel cell power systems provides an excellent case study illustrating the challenges of bringing a new technology to a commercial reality. EPRI's role in development of this novel power generation option has involved exploratory and applied research; component engineering development and full-scale demonstration projects; definition of market opportunities and identifying fuel cell products which satisfy those opportunities. The Institute is also playing an important role in bridging the barriers of commercialization by bringing together suppliers and buyers. EPRI is also enhancing the technical awareness of fuel cells internationally. Collaboration with foreign developers, information exchange and monitoring worldwide development efforts is becoming a more important element in the Institute's R&D planning. EPRI's past and current activities in developing and commercializing fuel cell power systems for electric utility applications are reviewed here.

Near-commercial — low temperature fuel cell systems

EPRI's fuel cell program has been in place since the Institute was founded; cumulative R&D expenditures to date have been approximately $70 million dollars. Because low temperature fuel cells of the phosphoric acid type (PAFC) were closest to commercial reality, emphasis during the past 10 years has been placed on developing a commercially viable phosphoric acid fuel cell power module; suitable for dispersed siting applications in large urban or municipal regions.

Research efforts in the early 1970s focused on fundamental work in understanding electrochemical reactions, improving the performance of electrocatalysts and identifying new electrolytes for use in PAFCs. Since the preferred fuel for early market power cells will be natural gas and possibly liquid fuels ranging from naphtha to heavier fuel oils, early research also focused on steam-reforming catalysts, fuel processing system development and steam-reformer reactor design so that fuel cells could effectively use the feedstocks.

After a successful verification of a 1 MW pilot plant in 1977 by United Technologies Corporation (UTC) and nine electric utilities, it became clear that further verification of the fuel cells operational, environmental and siting claims would be required before the industry would implement this

Fig. 2. 4.5 MW demonstrator at New York City site.

new type of power generation option in their systems. EPRI, together with ERDA [now the U.S. Dept. of Energy (DOE)], funded a multi-million dollar program with UTC to design, manufacture and test a 4.5 MW fuel cell power plant demonstrator in downtown New York City, Fig. 2. EPRI believed it was critical that this demonstrator be connected to the utility grid and operated by utility personnel. The demonstrator was also to provide a check on the fuel cell's operational features, including heat rate, power quality, transient response and start-up and shutdown characteristics — all necessary prerequisites for electric utility acceptance.

In 1980, Japan's Tokyo Electric Power Co. (TEPCO) also announced that it would install and test its own fuel cell demonstrator — a twin of the New York unit. This began a collaborative relationship between TEPCO and EPRI that continues today. Results from both the New York and the TPECO demonstration units were published [1, 2] and EPRI was able to identify many of the lessons learned and design deficiencies in the demonstration units and address these issues in ongoing development projects at UTC, where development of a market entry PAFC unit was proceeding in parallel to the pilot-plant demonstration projects.

During the 4.5 MW demonstration time period, EPRI recognized that the cost and complexity issues of PAFC systems and durability issues of fuel cell stacks would have to be addressed for market acceptance. EPRI, together with DOE, sponsored development of an improved fuel cell stack design and power plant system. EPRI focused on improving and simplifying the overall system design and ensuring that a fuel cell power plant, designed to utility standards, would meet performance, durability and operational needs the electric utility industry required. DOE focused on stack scale-up, manufacturing and cost reduction.

Results from EPRI sponsored work in the mid-1980s included: definition and preliminary design of a much improved 11 MW power plant system (Table 1, Fig. 3); technology development of key components in the system (reformer, inverter, controller) and verification of performance and durability of those components. PAFC stack manufacturing cost assessments and power plant capital cost estimates for near-commercial and advanced PAFC power plants were also developed. These activities contributed to the first pre-commercial offering of fuel cell power modules to the electric utility industry in 1986 [3 - 7]. TEPCO is currently constructing an 11 MW PAFC power plant of this design which is expected to be operational by 1991.

In the early 1980s, EPRI also sponsored Westinghouse Electric Corporation in the development of an air-cooled fuel cell system design. Emphasis again was placed on cost, performance and durability. A 7.5 MW PAFC

TABLE 1

Characteristics of improved pre-commercial fuel cell *vs.* 4.5 MW demonstrator design

	Demonstrator	11 MW pre-commercial configuration
Module size (MW)	4.5	11
Heat rate (Btu/kW h) HHV	9300	8300
Power range (%)	25 - 100	30 - 100
Plant operating pressure (psia)	50	120
Fuel cell active area (ft^2)	3.7	10
Plant foot print (acres)	0.8	0.8 - 1.0
Startup time (h)	4	6
Emissions (lb/million Btu)		
NOx	0.02	0.006
SOx	0.00003	0.0004
Relative installed cost ($/kW)		
(1987 $)		
(a) custom built, first-of-a-kind	5500	
(b) based on 3 units produced		3600
(c) 100 - 300 MW/year production		1550

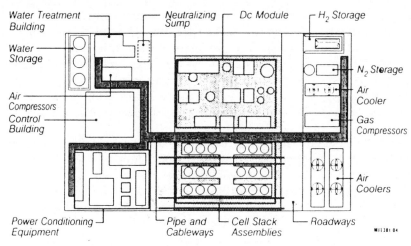

Fig. 3. Standard physical arrangement of 11 MW pre-commercial unit.

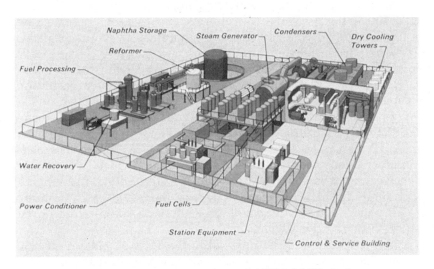

Fig. 4. Conceptual design of a Westinghouse PAFC 7.5 MW plant.

reference system was defined (Fig. 4), including a 1.5 MW pilot-plant demonstrator. Through an EPRI sponsored project at Westinghouse, a modular, high-efficient steam reformer was also developed and demonstrated at the 1.25 MW scale [8, 9]. This reformer technology is now available on a commercial basis from Haldor Topsoe, Inc. and is being applied to PAFC designs as small as 200 kW to 1000 kW.

As the phosphoric acid fuel cells became technically ready for the utility market in the mid 1980s, EPRI's role turned to commercialization. The Fuel Cell Users Group of The Electric Utility Industry (FCUG) was organized under EPRI leadership. Composed of over 55 utilities, the FCUG represented both potential early purchasers and technology advocates and was the first organized utility group to begin interacting with fuel cell developers. Through the vehicle of the FCUG, EPRI conducted several fuel cell market studies and application guides to assist developers in understanding market needs and to aid utility system planners in conducting accurate technology assessments and system expansion planning [10 - 13]. These studies identified the potential initial market opportunities for fuel cells and the necessary 'market entry' capital costs for electric utility deployment.

A study titled "EPRI Roles in Commercialization of Fuel Cells" showed that the most effective use of EPRI's R&D funds for PAFC commercialization was to assist in underwriting the initial high cost of the demonstration unit. And today, regardless of the type of fuel cell technology, underwriting a portion of the initial plant cost, or minimizing utility risk in other ways, continues to be an important role for EPRI in implementing fuel cell technology on utility systems.

Through EPRI leadership, the idea of forming utility consortiums to share the risks and rewards in commercializing fuel cells was also first introduced to the industry. While this approach worked in implementing the world's first integrated coal gasification — combined cycle power plant, U.S. utilities have been reluctant to demonstrate PAFC type fuel cells because of current market conditions and the potentially high risks.

Nevertheless, EPRI is monitoring large scale PAFC demonstration projects in Japan and Europe. Through an agreement with TEPCO, operational data and key results from TEPCO's 11 MW fuel cell plant will be made available to EPRI member utilities as well as limited access to the 11 MW site during operational testing. In return, EPRI is providing to TEPCO an experimental computer aided training and diagnostic aid which will enable TEPCO's operators to trouble-shoot operational faults during start-up and routine operation [14, 15]. This technology was pioneered by EPRI using system testability and failure mode and fault analysis to optimize diagnostic instrumentation during system design.

Through a collaborative agreement with the City of Milan's electric utility, Azienda energetica municipale, EPRI is also monitoring a 1 MW PAFC pilot-plant demonstration which will be sited on the municipal system.

Developmental — high temperature fuel cell systems

High temperature molten carbonate and solid oxide fuel cell systems offer the highest conceivable efficiency and lowest emission coal-based power generation technology known (Table 2). EPRI's long term goals are to couple advanced high temperature fuel cells with coal gasification technology (Fig. 5). However, before advanced fuel cells can be accepted by utilities to operate on coal gas, they must first be accepted using natural gas. Hence, EPRI's near term role in developing both advanced molten carbonate and solid oxide systems is focused on high value market entry products which are attractive and economical using natural gas.

TABLE 2

Design targets for integrated coal gasification/molten carbonate fuel cell systems

Characteristic	Design target
Application	Electric utility, baseload
Unit rating (MW)	200 - 250 module
Heat rate (Btu/kW h) HHV	6800
Fuel flexibility	all coals and natural gas
Emissions (lbs/MW h)	
NOx	0.1
SOx	0.005
particulates	trace
Installed cost ($/kW)	1200

Fig. 5. Conceptual integrated coal gasification/molten carbonate fuel cell power plant.

In the early 1980s, research sponsored by EPRI pioneered the development of the 'direct fuel cell'. In this concept, natural gas fuel (or methanol) is converted to a hydrogen rich gas on the anode surface of a molten carbonate fuel cell; the hydrogen produced is rapidly consumed electrochemically and converted into d.c. power. A fuel cell power system based on this concept is extremely efficient (60%+), simple in design since no external fuel processor is needed, and has attractive installed capital cost (Fig. 6).

Our approach here has included defining a commercially attractive system and using the system results to establish performance, scale and cost requirements of fuel cell stacks [16, 17]. In developing fuel cell stacks, the key thrust is to impose design-to-cost requirements consistent with commercial requirements. In this approach, fuel cell designs are analyzed/

54

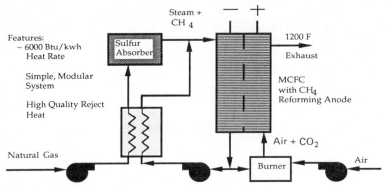

Features:
~ 6000 Btu/kwh
Heat Rate

Simple, Modular
System

High Quality Reject
Heat

Natural Gas

Steam +
CH$_4$

Sulfur
Absorber

1200 F

Exhaust

MCFC
with CH$_4$
Reforming Anode

Air + CO$_2$

Burner

Air

Fig. 6. Internal reforming molten carbonate system.

optimized to cost requirements before rigorous fabrication and testing programs are initiated. This minimizes development costs and insures fuel cell stacks will be technically acceptable as well as economically viable.

The development of the 'direct fuel cell' concept into a commercially acceptable dispersed generator is the current focus of EPRI's fuel cell research program. A completely integrated 100 kW pilot-plant is currently under construction. This unit will be installed at Pacific Gas & Electric's R&D facility near San Francisco, CA. Integrated system tests will be conducted by PG & E by the fall of 1990. This project is an important milestone in moving towards a 2 MW, 6500 Btu/kW h heat rate dispersed generator and smaller market entry products, Table 3, Fig. 7.

EPRI's strategy in developing advanced fuel cells relies heavily on the overall U.S. National Fuel Cell Program. In that program, EPRI, DOE and the Gas Research Institute (GRI) collaborate to prioritize research issues and coordinate areas of research to eliminate duplication. For example, EPRI and GRI are currently sponsoring development of early market, advanced fuel cell systems, and DOE is supporting research which will enable advanced fuel cells to operate on a coal gas. This year's coordinated EPRI and DOE efforts involve assessing MCFC and SOFC system integration,

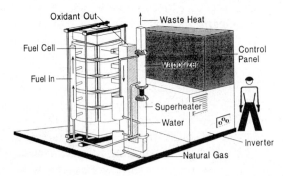

Fig. 7. Conceptual layout of a 100 kW MCFC market entry unit.

TABLE 3

Market entry MCFC system characteristics

Design parameter	Applications	
	Commercial on-site	Dispersed generator
Size, net power output (kW)	200 - 500	2000 - 5000
Electrical efficiency (%) HHV	50	55 - 57
Fuel	natural gas	natural gas
Site area (ft^2)	500 - 700	4000
Water supply make-up	0	0
Modularity	pallet assembled	pallet assembled
Emissions (lb/MW h)		
NOx	< 0.01	0
SOx	trace	< 0.01
CO	0	0
CO$_2$	820	820
hydrocarbons	0	0
particulates	0	0
Design life		
plant (years)	20	20
fuel cell stacks (h)	25000	25000
Availability (%)	85	85
Installed cost target ($/kW)	< 1500	< 1500

performance and capital cost requirements. Results will aid in guiding research programs.

Development and evaluation of solid oxide fuel cell (SOFC) components and system concepts has only recently been sponsored. Development and evaluation of a new planar SOFC concept was performed, and currently EPRI is funding research examining alternative SOFC design concepts. Over the next few years, EPRI will assess the feasibility of tubular SOFC systems using coal and/or natural gas. Through a collaborative agreement with the New Energy and Industrial Development Organization (NEDO) in Japan, EPRI is directing research on their behalf which will assess the technical and economic merits of large-scale (300 MW) and small (15 - 30 MW) SOFC system concepts.

Challenges of commercialization

To expedite the commercialization of fuel cells, EPRI has had to focus more on quantifying the market for fuel cells and understanding current 'market driven' utility needs. The efforts in this area have increased recently through working closely with the American Public Power Association (APPA). A recent market assessment, funded together by EPRI and APPA, identified the magnitude of the early market potential for fuel cells on public power systems, capital cost thresholds, and optimal unit size considerations. The 'likely early market' for fuel cells in public power is ~12 000 to 14 000 MW over the 1996 to 2010 time period; unit sizes ranging from 3 MW to 10 MW would be attractive; and installed capital costs near or below $1000/kW (1986 $) will be necessary to achieve a sizable mature market in public power [18].

The public power market is a potentially attractive early market 'niche' for near-commercial fuel cells. APPA's 'Notice of Market Opportunity' generated considerable interest from fuel cell developers in this market [19]. Currently, EPRI is assisting APPA utilities by providing technical expertise in reviewing vendor product offerings, business plans and overall approach to commercialization. The goal is to identify one or more commercialization approaches for a broad range of U.S. utilities by early next year.

Collaborative research

While certain U.S. utilities are developing fuel cell commercialization strategies with vendors, the majority of the industry (for numerous economic or market reasons) are waiting for the fuel cell to become commercial in the late 1990s.

Utilities in Japan and Europe are moving much faster in deploying precommercial fuel cell generators in their systems. In that regard, EPRI has established collaborative information exchange agreements with these utilities and foreign government funding organizations. The lessons learned and information exchanged from collaborative projects is valuable in establishing R&D needs; operating and maintenance data from early units can quantify risk and assist other utilities in evaluating the technology. International collaboration can shorten development time, minimize R&D investments and increase business opportunities worldwide.

EPRI has recognized that a U.S. research and development program cannot ignore worldwide technological developments. This is particularly true in the fuel cell area. International collaboration can truly make this promising technology a reality.

References

1 Implementation of a 4.5 MW fuel cell demonstrator in an electric utility system, *Consolidated Edison Company of New York, Final Report*, June 1987.
2 *Operational Experiences with the Tokyo Electric Power Company's 4.5 MW Fuel Cell Demonstration Plant*, International Fuel Cells, Tokyo Electric Power Co., EPRI, *Power Eng.*, March 1987.
3 Description of a generic 11 MW fuel cell power plant for utility applications, *EPRI Report EM-3161*, Sept. 1983.
4 Fuel processor development for 11 MW fuel cell power plants, *EPRI Report EM-4123*, July 1985.

5 Fuel processor development for multimegawatt fuel cell power plants, *EPRI Report AP-5349*, Aug. 1987.

6 Capital cost assessment of phosphoric acid fuel cell power plants for electric utility applications, *EPRI Report AP-5608*, March 1988.

7 *Product Description of The PC23 Fuel Cell*, International Fuel Cells, South Windsor, CT, Jan. 1987.

8 Demonstration of a high-efficiency steam reformer for fuel cell power plant applications, *EPRI Report AP-5319*, Aug. 1987.

9 Endurance testing of a high-efficiency steam reformer for fuel cell plants, *EPRI Report AP-6071*, Oct. 1988.

10 Application of fuel cells on utility systems, *EPRI Report EM-3205*, Aug. 1983.

11 Systems planner's guide for evaluating phosphoric acid fuel cell power plants, *EPRI Report EM-3512*, July 1984.

12 The financial and strategic planning benefits of fuel cell power plants, *EPRI Report EM-4511*, April 1986.

13 EPRI roles in fuel cell commercialization, *EPRI Report AP-5137*, April 1987.

14 System testability and maintenance program for troubleshooting in fuel cell power plants, *EPRI Report AP-6017*, Sept. 1988.

15 An interactive computer aided troubleshooting system for fuel cell power plants, *EPRI Report GS*, to be published Dec. 1989.

16 Assessment of a 6500-Btu/kwh heat rate dispersed generator, *EPRI Report EM-3307*, Nov. 1983.

17 Parametric analysis of a 6500 Btu/kwh heat rate dispersed generator, *EPRI Report EM-4179*, Aug. 1985.

18 The market for fuel cell power plants within municipally-owned electric utilities, *EPRI Report GS-6692*, to be published Dec. 1989.

19 *Notice of Market Opportunity for Fuel Cells*, issued by the American Public Power Association, Washington, DC, Oct. 1988.

WESTINGHOUSE AIR-COOLED PAFC TECHNOLOGY

JEROME M. FERET

Westinghouse Electric Corporation, Advanced Energy Systems, NATD, Large, Pittsburgh, PA 15236-0864 (U.S.A.)

Introduction

Over the past 20 years, the Department of Energy (DOE), the Electric Power Research Institute, the Gas Research Institute, private industry and others in the U.S.A. have been pursuing the development of fuel cells for use in environmentally clean electric utility and industrial power plants. These power plants are expected to be in the 3 to 50 MW range.

Of the several types of fuel cells, the phosphoric acid fuel cell (PAFC) technology is the furthest developed, and thus most mature in terms of readiness for commercialization. The Westinghouse Electric Corporation entered into a licensing agreement with Energy Research Corporation relative to their air-cooled PAFC technology. Air rather than water was selected for cooling as this avoids the need to incorporate additional cooling paths into the stack with resultant added series resistance and corrosion-complications. As a result of this agreement, Westinghouse has been developing for over a decade this most promising and highly efficient alternative power generation technology option. Plans to bring this technology to the commercial marketplace, the power plant key features and fuel cell technology status are now examined.

Program overview

The Westinghouse PAFC program consists of two complementary but highly integrated programs. These programs are the Westinghouse sponsored Power Plant Program and the United States DOE sponsored PAFC Technology Development Program. Under the Power Plant Program, Westinghouse along with its other team members will design, build and operate demonstration and commercial systems for various utility and industrial plant applications. The cell technology development effort is being performed by a joint Westinghouse and Energy Research Corporation team under the DOE Morgantown Energy Technology Center Contract DE-AC21-82MC24223. The key objective of these programs is to commercialize the technology in the 1990s for the electric utility and industrial markets.

0378-7753/90/$3.50

Commercialization

The Westinghouse commercialization program is directed towards having standardized factory produced 3 and 13 MW power plants available for sale commercially by the mid-1990s. Figure 1 illustrates the commercialization plan which consists of three distinct but highly inter-related phases. These are the prototype demonstration, early commercialization and mature commercialization phases.

Acceptance of the commercial plants will depend largely on the level of confidence in performance and economic predictions. Therefore, Westinghouse intends to build a 3 MW prototype power plant as the next step in the commercialization chronology. As described later, the Westinghouse air-cooled PAFCs are ready for prototype operation under real load conditions.

Following this demonstration, the Westinghouse team will evaluate the performance of the PAFC modules along with other critical plant systems such as fuel processing and power conditions. In addition, updated plant system and component cost projections, plant economics and relevant market data will be analyzed. At this time, Westinghouse intends to construct a highly automated manufacturing facility by the end of 1994 to initially mass produce fuel cell modules. This facility, along with other cost reductions envisaged, is expected to result in an acceptable initial selling price for the plant.

The final step in the plan is the mature commercialization phase. This phase is expected to be completely market driven. It is projected that the

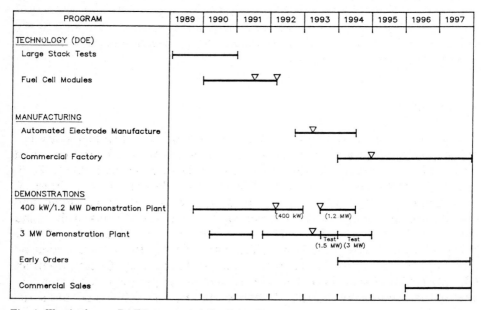

Fig. 1. Westinghouse PAFC commercialization plan.

early PAFC units produced in this phase may be somewhat higher priced relative to competitive electrical power generation technologies. This price, however, will be offset considerably by the value of the environmental acceptability and other well known benefits of fuel cell power plants particularly as more stringent siting and environmental regulations become law.

Plant design

Westinghouse has always placed a most important emphasis on the reliability and maintainability aspects of the power plant. This consideration, along with others, resulted in the choice of an air-cooled PAFC design along with a modular power plant approach. Multiple units of the 3 and 13 MW plant sizes may be grouped at the same site or dispersed areas to cover the expected power plant range of 3 to 50 MW or larger.

The Westinghouse 3 MW power plant for an all-electric application is shown in Fig. 2. This plant is designed to operate on a variety of fuels such as natural gas, methanol, light distillates, coal gas and other light hydrocarbons. For an electric utility application, the plant will operate as an intermediate or base loaded plant. These characteristics are anticipated to make the plant attractive for a broad range of utility and industrial applications.

The plant can be sited on about 4000 square meters (1.0 acre) of land and consists of a number of major systems. These include the Fuel Processing Power Conditioning, Rotating Equipment, Steam Generation,

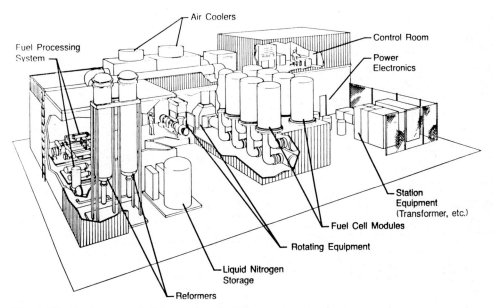

Fig. 2. Westinghouse prototype power plant.

Instrumentation and Control, Balance-of-Plant, and Fuel Systems [1]. The nucleus of the plant is the fuel cell system. This system contains 8 fuel cell modules that are appropriately coupled electrically while thermally in parallel. Each of these modules has a 375 kW nominal rated electric output at beginning-of-use. The air cooled fuel cells are inherently simple and reliable, as the power plant process flow schematic in Fig. 3 illustrates. The Westinghouse 375 kW module is the basic power plant building block. This module is shown in Fig. 4. Four series connected stacks, each containing approximately 450 cells, comprise the Westinghouse 375 kW module. These four cell stacks are housed within a vessel that contains the pressurized fuel and air process gasses.

Initially, these modules are planned to be operated at the following beginning-of-use conditions:

- Pressure   80 psia
- Temperature   205 °C
- Current density   267 mA/cm²
- Fuel/air utilization   83/60%
- Cell voltage   720 mV

As shown in Fig. 3, the Fuel Processing System supplies the hydrogen rich fuel gas to the anode side of the PAFC. Cooling air is supplied to the pressure vessel and flows radially inward through cooling passages provided between groups of six cells to a plenum formed by the four cell stacks. A

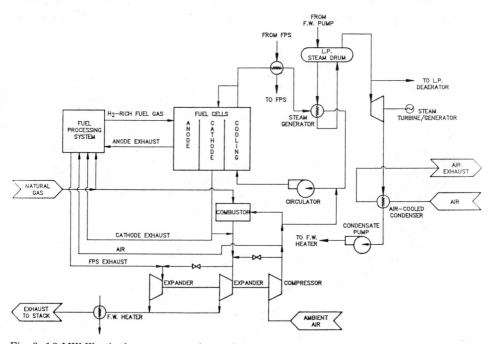

Fig. 3. 13 MW Westinghouse power plant schematic.

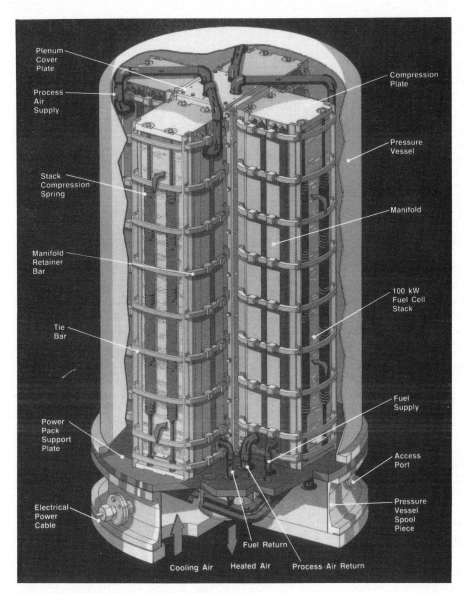

Fig. 4. Westinghouse 375 kW fuel cell module.

small fraction of the heated cooling air is extracted and directed to the cathode side of the cell. Energy to maintain the flow of fuel cell cooling air is supplied by a pressurized circulator in the Rotating Equipment System. The Steam Generation System utilizes the fuel cell waste heat to generate additional electric power. The Power Conditioning System interfaces with the fuel cell modules through consolidation circuits. These circuits control the module currents, which are combined at a common d.c. bus. The

TABLE 1

PAFC plant characteristics

Rated power	13.8 MW
Heat rate (LHV)	6900 Btu/kW h
(HHV)	7674 Btu/kW h
Availability	90%
Plant design life	30 years
Ramp rate	1 MW/s
Operation	unattended remote dispatch
NOx emissions	5 ppm at 15% O_2

projected top level performance characteristics of the Westinghouse 13 MW power plant are summarized in Table 1.

Westinghouse firmly believes that fuel cells offer the capability for self-generated power and local control of the power supply. This is rapidly becoming very important as public power, for example, experiences the impacts of deregulation, generation and transmission capacity access, and mounting environmental problems. The American Public Power Association has recently invited fuel cell manufacturers to form a partnership to commercialize multi-megawatt PAFC plants.

Fuel cell technology

The Westinghouse commercialization plan success is highly dependent upon the U.S. DOE sponsored Technology Development Program to provide a PAFC technology that meets certain performance and economic objectives. The early PAFC technology objectives defined to achieve the plant established goals include: (1) average beginning-of-use performance of 690 mV at 267 mA/cm^2, 190 °C, 4.7 atmospheres, 83% hydrogen utilization using reformed natural gas and 50% oxidant utilization using air; (2) performance stability consistent with a voltage loss of less than 8 mV/1000 h; and (3) a 375 kW module that can be manufactured for about \$2600/kW without employing mass production techniques. Modest performance and performance stability improvements to 720 mV and 2 to 4 mV/1000 h, respectively, are needed to achieve the mature commercial power plant goals.

Extensive experiments, subscale cell screening tests, cell materials and components, characterizations, and stack tests have been performed over the past decade. These various efforts resulted in the selection of a baseline cell technology and associated stack design to achieve the performance objectives.

Cell technology selections were made. These selections included the design configuration, materials of construction, manufacturing process(es), subassembly and assembly techniques. In excess of 1 000 000 subscale cell and 27 000 stack test hours were accumulated to assist in the selection

TABLE 2

Cell baseline technology

Catalyst	platinum on carbon
Catalyst layer	Energy Research Corporation rolled configuration
Electrode support	wetproofed carbon paper
Electrodes	0.5 mg/cm^2 platinum for cathode and 0.25 mg/cm^2 for anode
Matrix	silicon carbide–carbon layer composite
Bipolar plates	heat treated graphite–phenolic resin composite
Seals	three piece teflon
Acid make-up	four corner feed

TABLE 3

Small stack performance

Stack	Cell voltage (mV)
W010-22	694
W010-23	699
W010-24	702
W010-25	689

process. The specific components involved and selections made are defined in Table 2.

Four essentially identical ten cell stacks were constructed. Two of these stacks were tested at the rated operating conditions for in excess of 5000 h. The third was tested for over 16 000 h which at this time represents the world's endurance record for a pressurized stack of PAFCs.

The beginning-of-life performance for each of these stacks is shown in Table 3. As can be noted, the beginning-of-life performance for each of the stacks essentially met or exceeded the 690 mV goal. The average performance for these stacks is 696 mV/cell with a standard deviation of 5 mV.

The performance decay for one of these stacks, namely Stack W010-22, is presented in Fig. 5. As noted, the 8 mV/1000 h voltage decay goal is nearly achieved with 8.3 mV/1000 h obtained over 16 000 h of testing. The voltage decay goal was achieved early in life for the other two stacks (≈1500 h) while a decay of about −12 mV/1000 h was observed over 5000 test hours.

With this repeated and quite satisfactory performance, development emphasis was shifted to the non-repeating components associated with the larger size stacks required for the plant fuel cells modules. Four 152-cell stacks rated at 32 kW were constructed. Each of these stacks exhibiting prototypic characteristics of the module stacks was tested individually and in appropriately combined 64 and 96 kW configurations. The cell tech-

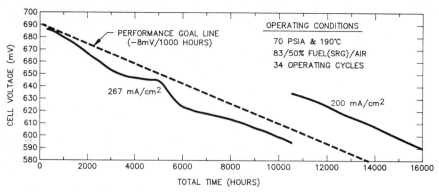

Fig. 5. Stack W010-22 performance stability.

TABLE 4

Large stack performance

Stack	Cell voltage (mV)
W152-01	694
W152-02	690
W152-03	688
W152-04	692

nology used in each of these stacks essentially duplicated that used in the earlier discussed ten cell stacks.

The beginning-of-life performance for each of these stacks is shown in Table 4. As shown, the beginning-of-life performance for each of these stacks nearly met or exceeded the 690 mV goal. The average performance for these stacks which represent over 600 individual cells is 691 mV. Furthermore, an average variance of 16 mV was achieved for some 100 six cell groups involved which represent the basic cell building block for stacks.

The performance decay for one of these stacks, namely Stack W152-04 is presented in Fig. 6. The overall voltage decay rate for this stack is 9.6 mV/1000 h. As shown, this rate of cell voltage loss was driven by an abnormally high rate of −23 mV/1000 h over the first 1500 h of testing.

This rate does not reflect the cell/stack technology but rather was incurred as a result of several unfortunate facility upsets and an operator error. This is supported by the achievement of 8.3 mV/1000 h voltage loss over the next 1500 h of testing.

In addition to having met or exceeded the initially established performance and technology scale-up goals, needed improvements in selected areas of the cell and stack were identified. The challenges to improve the performance included: a lower cell resistance; higher operating temperature; improved catalyst activity and design configuration; and process control improvements. To reduce the cell voltage loss involved: more corrosion

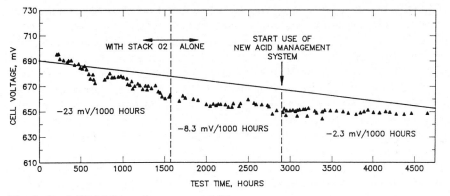

Fig. 6. Stack W152-04 performance stability.

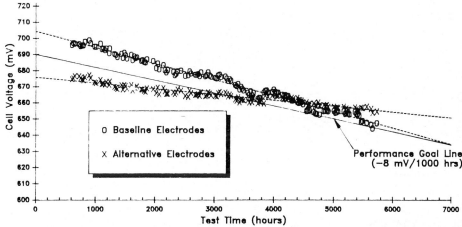

Fig. 7. Stack E010-09R.

resistant catalyst supports; a better electrolyte management system; and lower platinum or activity loss. Improving the module cost involved: single unit module stacks; improved matrix; electrode integral cell seal; and alternative electrode and plate manufacturing processes. During the past year, the thrust of the technical effort was directed towards pursuing the various solutions identified for each of these challenges.

After having completed appropriate screening tests, a more corrosion resistance catalyst support material was selected. A 10-cell stack was constructed which is comprised of five cathode electrodes containing an alternative catalyst carbon support and the balance the baseline support material. The performance of this stack is provided in Fig. 7. As shown, the performance stability of the alternative catalyst support cells is a factor of two less than the baseline cells, or −4.6 *versus* −10.9 mV per 1000 h. This improved voltage decay rate, however, is at the expense of a 21 mV loss in beginning-of-use performance or 698 *versus* 677 mV. The respective rates of voltage decay have been consistent for nearly 9000 test hours.

Another means of reducing the voltage decay rate involves an improved acid management system. This improved system was first used in 10-cell Stack W010-27 (Fig. 8) and then the 152-cell Stack W152-04 (Fig. 6). Similar encouraging results have since been demonstrated in 10-cell Stack W010-29 (Fig. 9) and the 100 kW Stack W446-01. As can be seen, the voltage decay rate was improved in each of these stacks to less than 4 mV/ 1000 h. This improvement was achieved using the baseline catalyst support material. In addition, Stack W010-29 contained a new matrix structure and configuration. The MAT-1 carbon layer was replaced with a vendor-supplied thin carbon layer.

Several efforts are underway to increase the beginning-of-use performance by 15–20 mV minimum. These include a thinner matrix, use of an alloy catalyst, higher platinum loadings and higher temperature operation. Initial test results indicate that a 25 mV improvement can be achieved [2, 3].

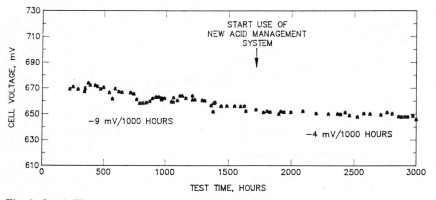

Fig. 8. Stack W010-27 affect on new acid management system.

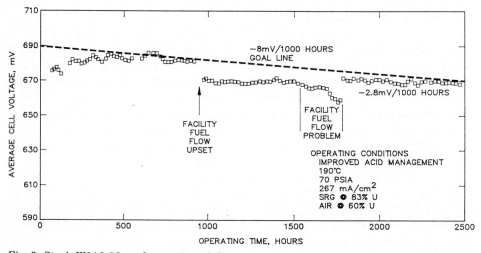

Fig. 9. Stack W010-29 performance stability.

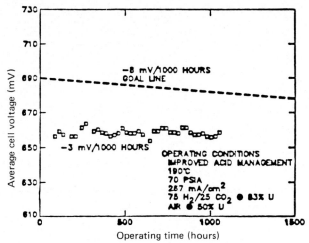

Fig. 10. Stack W446-01 performance stability.

Many areas are under investigation to meet the module cost objective of $2600/kW without employing mass production techniques. These encompass the cell materials, manufacturing processes and assembly methods as well as the cell/module design. The results of developments in several key areas are in the process of being finalized. These involve: (1) use of a thinner electrode support layer, (2) replacement of the MAT-1 carbon layer with a vendor-supplied product, and (3) integral electrode edge seal.

A most important element of our cost improvement plan involves the fabrication and testing of a single unit 446-cell stack. This stack is nominally rated at 100 kW. Its features are the most prototypic of the four stacks that will be used in the projects' end product, namely the 375 kW module.

The stack performance in all aspects was a notable success. The procedures, tooling and fixturing developed to assemble this single unit eight foot tall stack were all trouble-free. Perhaps, the most important achievement involves the leakage free performance of the process gasses' manifold seal joint. The overall voltage decay rate for this stack is less than 4 mV/1000 h for nearly 1600 h of operation, as shown in Fig. 10.

In summary, the technology is sufficiently in hand that proven solutions to the remaining challenges are being developed in a timely fashion. This in turn will thus allow their demonstration at the module level as planned.

References

1 D. Newby, Westinghouse air-cooled phosphoric acid fuel cell program, *EC-Italian Fuel Cell Workshop, Taormina, Italy, 1987.*
2 Various monthly and quarterly technical progress narratives pertaining to DOE Contract DE-AC21-82MC24223.
3 J. M. Feret, *Westinghouse Air-Cooled PAFC Status, Accomplishments and Issues,* Fuel Cell Contractors Review, Morgantown, WV, 1989.

Journal of Power Sources, 29 (1990) 71 - 75

TECHNOLOGY DEVELOPMENT AND MARKET INTRODUCTION OF PAFC SYSTEMS

L. J. M. J. BLOMEN* and M. N. MUGERWA

Kinetics Technology International Group B.V., 26 Bredewater, P.O. Box 86, 2700 AB Zoetermeer (The Netherlands)

Introduction

New power generating technologies need to be clean, highly efficient (also at partial load), highly reliable and competitively priced. Moreover, they need to be constructed as modules to enable step-by-step capacity increases at an acceptable cost level, and increase planning flexibility as a consequence. They must have low maintenance and low noise levels. Phosphoric acid fuel cell system technology to date has been shown to meet most of these requirements, except for two major issues: the reliability of power plants built and demonstrated to date has been inadequate (hence high maintenance requirements), and cost price levels have not yet been reduced to commercially competitive levels.

Reliability

The reliability of the current generation of fuel cell power plants is one of the most critical issues facing the commercialisation of this technology today. As far as most potential users are concerned, reduced reliability means increased maintenance costs and an increased dependence upon more costly back-up supplies of electrical energy or a need for redundant power supply capacity (extra investment). In fact, in most instances, diminished reliability makes the application of such technology decidedly unattractive, this despite all the advantages attributed to fuel cell systems. Most of the world's fuel cell demonstration programmes have managed to highlight those well-known attributes of fuel cell technology, namely, high efficiency, high part load efficiency, environmental friendliness and modular characteristics. However, many of these programmes have been unable to initiate the round of expected commercialisation of this technology by singularly failing to demonstrate that fuel cell systems can indeed be highly reliable.

Analyses of the reasons for the reduced reliability of many fuel cell power plants in the U.S. and Japan have shown that by far the vast majority of forced shutdowns (>95%) have been caused by the failure of the balance

---

*Author to whom correspondence should be addressed.

0378-7753/90/$3.50

of plant components. These stoppages have involved electrical and electronics components (including sensors and cabling), mechanical controls, leakages, reformer malfunctions and rotating equipment shortcomings. In short, those elements that represent established and well-known technologies which are of course in stark contrast with relatively high availability of the 'new' fuel cell stack technology itself.

Modern hydrogen plants, that are in widespread use in the electronics, food, metallurgical and petrochemicals industries as well as in refineries, require constant availability, 24 h per day, over 8400 h per year, without any unexpected shutdowns. Industry uses hydrogen like individuals use tap water or electricity from the grid. Reliability of plant operation is a matter of design philosophy, engineering procedures and quality control. In fuel cell power plants, standard components should be used that have proven operation in existing hydrogen plants. This fact favours larger capacity fuel cell plants, where many more proven components can be used than in small capacity fuel cell plants (<100 kW). In the latter case, dedicated development of many components complicates the matter, though once proven, mass production can more quickly reduce the costs. As an established engineering contractor with extensive experience in all aspects of industrial hydrogen plant technology, where reliable on-stream operation is an economic necessity, KTI is in an excellent position to utilise its knowhow for the benefit of fuel cell system commercialisation.

Economics

In addition to the necessity to demonstrate the reliability of fuel cell power plants, the other major issue currently facing fuel cell commercialisation is that of economics. Broadly speaking, the economics of electric power generating systems can be segregated into fuel costs, operating and maintenance (O&M) costs and capital costs. Current phosphoric acid fuel cell power plant designs can achieve efficiencies in excess of 40%, and it is expected that in the near future (next three years) net system efficiencies will be dependent upon improvements in fuel cell stack, d.c./a.c. inverter and rotating equipment performance. These high efficiencies will ensure that the fuel costs of present and future fuel cell power plants are, and will remain, competitive.

Once the important objective of increased reliability of fuel cell power plant systems is achieved, it is realistic to expect significant reductions in O&M costs. Significant O&M cost reductions will also be achieved through decreasing capital costs requirements, e.g. fuel cell stack replacement costs after 40 000 hours of operation. In addition, as technical and operating experience with fuel cell power plants grows, important reductions in O&M costs will be achieved through increased automation of the systems.

The most important improvements in fuel cell system economics will be achieved through reductions in installed capital costs. These cost reductions will only be achieved through increased production volumes of fuel cell systems. The serial production phase of fuel cell system commercialisation will be attained once market acceptance has been achieved. This will only be possible once the potential high reliability of fuel cell systems has been finally demonstrated. At KTI, based upon optimised flowsheet designs, detailed cost estimates for fuel cell power plants have been generated. The capacities of fuel cell power plants studied range from multi-kW through to multi-MW systems. In this way, the economy of scale could be compared with the effect of increasing serial production on the investment costs, and subsequently on the cost of electricity (COE). In order to achieve the reductions in cost necessary to make fuel cell systems competitive with conventional and also with other advanced power generation sources, the fuel cell power plants will have to be serially produced in modules, especially at low capacity.

Serial production has the advantage of reduction of prefabrication labour and overheads, long-term negotiated supply contracts, reduced project contingency, reduced control system costs (*e.g.* through the development of microprocessor control), and the reduction in planning and software requirements. Modular units with a maximum amount of prefabrication will also enable significant reductions in on-site installation costs. In addition, of course, fuel cell manufacturers themselves will also enjoy the benefits of automised serial production of fuel cell stacks. For example, recent studies at KTI have estimated that for a 3 MW_e fuel cell power plant the installed costs can be reduced from \$2800/kW for a first series of 5 units, to below \$1000/kW with serial production.

Sensitivity analyses of all important system parameters have shown areas where relative improvements to reduce the ultimate cost of electricity (COE) are most effective. This means that on the basis of system economics, system designs can be modified to yield any combination of better performance, lower cost, cleaner exhaust gas, etc. This generates feedback that enables design simplification, reformer performance improvements, changes in burner configuration, and can even set targets for fuel cells R&D itself! For example, it has been shown that the current density of phosphoric acid fuel cells has a different optimum value when targeting for a minimum COE for the total system, to when targeting for a minimum fuel cell stack cost per kW_e produced. Most fuel cell manufacturers will schedule their developments according to the latter criterion (equivalent to maximising the number of kW_e per m^2 of cell surface area), whereas a more economic product will result when designing the fuel cell for a global optimum COE instead of the maximum current density.

Technology development

KTI's hydrogen and fuel cell development programmes run essentially concurrently with one another, and include the following activities:

- The development of fully automatic, advanced mini-hydrogen plants
- Burner and reformer development and testing
- Two 25 kW breadboard units in the U.S. and Italy
- A European demonstration program (see Fig. 1) in which a few palette-mounted, completely automated 25 kW_e (a.c.) fuel cell power plants are to be demonstrated, the first of which is currently being started-up in Holland
- A combined electricity (80 kW_e) and pure hydrogen producing unit for Solar-Wasserstoff-Bayern sold on a commercial basis
- Designs and proposals for larger capacity units ranging from multi-hundred kW_e units to several MW_e capacity
- Dynamic simulation of fuel cell power plants with which partial load behaviour, load following characteristics, and dynamic responses to changes in process conditions are being extensively investigated
- High performance rotating equipment studies and selection
- Special efforts to reduce the number of unit operations and equipment items in fuel cell power plants

Fig. 1. After installation of the 25 kW demo unit at TU Delft, KTI personnel are making a final inspection of the unit. (The fuel cell stack, supplied by Fuji Electric Corp. in Japan, is located on the right-hand side of the skid and the fuel processor on the left-hand side of the skid.)

- Simplification of plant process control requirements
- Several optimisation techniques, including pressurised combustion, gas separation options (CO_2-removal, air enrichment, hydrogen purification, inerts separation)
- Modular construction
- Cost reduction studies
- Fuel flexibility

Market introduction

As discussed earlier, during the market introduction phase of fuel cell technology it is the reliability and cost issues that come to the forefront. KTI's approach to aiding market introduction of this technology is to ensure a reasonable risk distribution between the potential clients (especially in the initial market niches) and the fuel cell system suppliers. This will be achieved by undertaking the following measures:

- A willingness to undertake negotiated shared responsibility
- The possibility of guarantee provisions
- No major development effort involved in process design, *i.e.* making a maximum use of advanced hydrogen plant technology. This approach implies a minimum development effort with peripheral equipment and a concentration on the fuel cell stack integration, rotating equipment and the electrical requirements of the power plant
- The use of the feedback mechanism from demonstration plants to affect design, performance and cost improvements of the early production units
- Cost reduction studies
- Implementation of cost reduction through the use of existing manufacturing facilities for the production of plant modules
- A continuing assessment of market needs and related technology for developments

In conclusion, KTI is intending to direct its hydrogen and fuel cell development efforts towards addressing the most critical issues facing fuel cell system commercialisation today, namely those of reliability and cost.

Journal of Power Sources, 29 (1990) 77 - 83   77

THE MOLTEN CARBONATE FUEL CELL PROGRAMME IN THE NETHERLANDS

PIETER VAN DIJKUM*

Netherlands Agency for Energy and the Environment, NOVEM, Leidseveer 35, P.O. Box 8242, 3503 RE Utrecht (The Netherlands)

KLAAS JOON

Netherlands Energy Research Foundation, Postbus 1, 1750 ZO Petten (The Netherlands)

Introduction

NOVEM started to manage the Dutch Fuel Cell Programme in 1986. The programme exclusively supported research on molten carbonate fuel cells (MCFC) and aimed at an international research position comparable with those in the U.S.A. and in Japan. Also the interest of an industrial partner had to be raised.

Focus points in the programme were material research in combination with fabrication techniques, the operation of a 1 kW MCFC stack in 1989 and international cooperation. ECN was the exclusive contractor with support of the Technical University of Delft. ECN agreed with IGT, Chicago, U.S.A. on a programme for support and education.

Hoogovens/ESTS joined the programme in Spring 1987. The research programme was intensified with development activities. A much tighter time schedule for upscaling the stack components was envisaged supported by system studies. Also two field experiments of 25 kW phosphorous acid fuel cell (PAFC) systems by KTI were partly financed.

In October 1987 the Ministry of Economic Affairs expressed again their interest in fuel cells on the basis that they:
- Are environmentally benign
- Require a high level of technological research in which electrochemistry and material technology play an important role
- Offer new prospects for Dutch industry
- Make use of fossil fuel, natural gas, in an efficient way

Hoogovens/ESTS unfortunately withdrew from the R&D programme in Fall 1988. Although the technical and economic prospects were acknowledged as being good, the MCFC technology did not fit into their core business and it would have required too much of their available R&D capacity.

---

*Author to whom correspondence should be addressed.

0378-7753/90/$3.50

NOVEM consequently temporized the developing activities. At this time the following time schedule for stack development is being aimed at:
- Second half of 1989 a second 1 kW stack
- 1990 a 2.5 kW stack
- 1991 a 10 kW stack

At this moment NOVEM and ECN are putting much effort into attracting successors of Hoogovens/ESTS for participation in a joint venture to produce and commercialize MCFC systems in the Netherlands. If we succeed in establishing a joint venture in the next months then the following development phases will be executed:
- Phase 1, period 1986 - 1991, research and development, 10 kW stack in 1991
- Phase 2, period 1991 - 1995, preparation of the commercial phase with supporting R&D activities
- Phase 3, after 1995, commercialization of large units (250 kW and more) with supporting development activities

NOVEM and ECN are optimistic in establishing the joint venture.

Up to this moment NOVEM has spent about Dfl. 28 mln. financial means of the Ministry of Economic Affairs and ECN about Dfl. 8.5 mln. Now two questions can be raised:
- Is the MCFC technology still promising from economic and environmental viewpoints?
- What is the R&D status in the Netherlands?

Economic and environmental prospects of MCFC systems

Two comprehensive studies about MCFC systems are available. The first is the much quoted KTI study, titled "Fuel Cell Systems", May 1988. The second study is the ESTS report "Process Design of a Pilot Plant for MCFC", March 1989 (in Dutch only). Both engineering studies focussed on the optimal component configuration for producing electricity with high efficiencies. The important components in an external reforming fuel cell system are:
- Fuel Processing equipment, including the reformer, and piping
- Fuel cell stack
- d.c.–a.c. convertor

Besides the engineering of the system, KTI also executed detailed cost studies. Two effects on cost reduction were included in their calculations, namely:
- Learning curve experience
- Upscaling of the MCFC plant size capacity

KTI defined different levels of production capacities (see Table 1). The upscaling of the system size ranged from 25 kW up to 100 MW systems.

Results of the calculations on the installed plant costs per kW are presented in Fig. 1. The cost reduction for a 250 kW system of 65% was calcu-

TABLE 1

Number of units to be produced of different fuel cell system capacities[a]

Capacity fuel cell	Number of units			
	First unit	First series	Low volume production	High volume production
	FU	FS	LVP	HVP
250 kW		5	20	200
3250 kW		5	20	100
100 MW	1	5	20	

[a]Source: KTI.

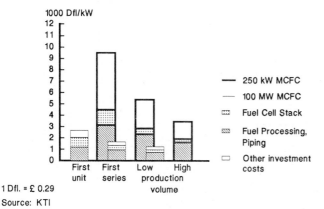

1 Dfl. = £ 0.29

Source: KTI

Fig. 1. Investment costs for MCFC systems (Dfl/kW) for different capacities and production levels.

lated when the 200th unit would be installed in comparison with the 5th unit of the first series. The dominant cost factor is still the fuel processing equipment including the piping. The cost percentage of the fuel cell stack ranges between 10 and 15%.

Even larger cost reductions are feasible by increasing the plant capacity from 250 kW up to 100 MW. Cost percentage of the fuel cell stack ranges in this case between 20 and 30%.

The system efficiencies must be estimated in order to calculate the price/performance ratios or in other words the cost of electricity production.

ESTS calculated the operating area of a MCFC stack of 25 MW. The system should preferably operate between 3 and 9.5 bar with an optimum of 7.5 bar with an electrical efficiency of 59% (LHV of natural gas). The results agree with KTI results although these results were calculated for different plant capacities (see Fig. 2).

The same effects of the cost reducing factors are clearly visible in the costs of electricity production at different production capacities and plant

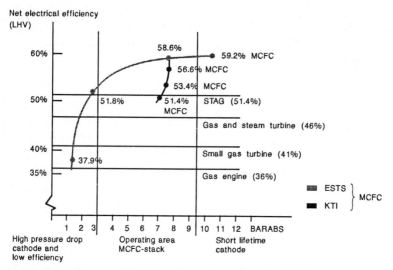

Fig. 2. Operating area MCFC stacks and net electrical efficiencies.

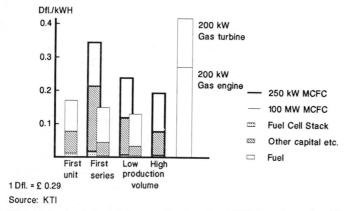

1 Dfl. = £ 0.29

Source: KTI

Fig. 3. Cost of electricity production for MCFC systems for different capacities and production levels.

sizes (see Fig. 3). At low production volume, the 20th 250 kW MCFC plant in a series of 20 could compete with small gas engines and gas turbines. The much larger 100 MW units could produce electricity at a cost price which is comparable with the cost price of a steam and gas turbine.

The characteristics of, for example, a 250 kW plant producing heat and electricity are good. The total system efficiency from the lower heating value of natural gas to useful energy is 80% of which 52% is the electricity generating part. The heat/power ratio is 0.55 (see Fig. 4).

The load following characteristics are good. A 40 kW PAFC unit supplying an electricity resistance heater for a sauna in a hotel in Tokyo follows for example immediately the variations in the demand between 17 and 34 kW.

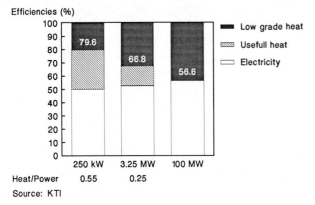

Fig. 4. MCFC system characteristics for coupled production of heat and electricity.

The NOx emissions are negligible when low NOx burners, or more preferably ceramic fiber burners in the reformer converting natural gas into hydrogen, are being used. The CO_2 emissions are relatively low because of the high efficiency of the technology and the possibility to recycle CO_2 from the anode off-gas to the cathode supply gas. SO_2 emissions are not present. The noise level is relatively low about 60 dB(A).

The conclusion is that the MCFC technology still offers the opportunity of a highly efficient, profitable and extremely benign fossil fuel conversion technology suitable for a broad range of (heat and) power applications. For the R&D programme it is necessary to research:

● The lifetime of the electrodes when stacks will be operated at 7.5 bar
● The effects of operating under pressure on internal reforming

Status of the research and development programme

About 50 people of several disciplines are working in the R&D field for MCFC systems using all kinds of test facilities and fabrication techniques for porous components and separator plates.

In the research part of the National Programme emphasis has been put on:

● Electrochemical research
● Improvement of state-of-the-art materials and new materials for the electrodes (conductive ceramics) and matrices
● Design and fabrication of separator plates
● Corrosion test with state-of-the-art materials for separator plates
● Improvement of the steps in the fabrication techniques such as powder processing, tape casting and sintering of improved state-of-the-art and new materials
● Testing of porous components under laboratory conditions (3 cm^2 and 3% use of hydrogen) and practical conditions (100 cm^2 respectively 1000 cm^2 and 75% use of hydrogen)

- Modelling of the ion migration through the electrolyte
- International cooperation

In the development part of the National Programme the attention is focussed on:

- Batch fabrication of the 1000 cm^2 porous components and separator plates and the design and engineering of scaling up the facilities to 4000 cm^2 (semi series production)
- Testing of the 1000 cm^2 components in the stackable cell test facility
- Testing the 1 kW stack (10 cell with an active area of 1000 cm^2 each) under atmospheric conditions
- Engineering of the 10 kW stack test facility up to 4000 cm^2 and 7 to 8 bar
- System studies
- International cooperation

The progress in the R&D programme can be illustrated by the results obtained in the test facilities:

- Laboratory cells of 3 cm^2 with improved state-of-the-art materials show performances in the range of 900 to 940 mV at a current density of 480 mA
- Bench scale cells of 100 cm^2 show voltages of 800 mV at 15 A
- The test with the subscale cell of 1000 cm^2 resulted in 775 mV at 150 A
- Subscale ministack test (two cells of 1000 cm^2 each in between three separator plates) resulted in 1450 mV at 150 A
- The best performance of the 1 kW stack was 5.81 V at 130 A

On the basis of these results the following conclusions could be drawn:

- The scaling up of the porous components from three to one thousand square centimeters has been successfully executed
- The operating parameters such as start-up procedures, holding force equipment, continuous calibrations etc. of the laboratory, bench scale and stackable cell test facilities are quite good
- The design of the bipolar separator plate has to be improved in order to boost the performance of the second 1 kW stack at the end of this year.

Some other highlights from the programme are:

- Promising candidates for alternative anode materials have been selected
- NiAl anode will probably be the standard anode in the future ECN programme
- Fabrication procedures and routes are improved
- The further need of a bubble barrier for preventing cracking of the matrix materials, is questionable
- The concept of the bipolar separator plate with integrated manifolds for the reactant gasses has proved to function
- The modelling activity is at this moment already a powerful instrument for supporting stack engineering activities

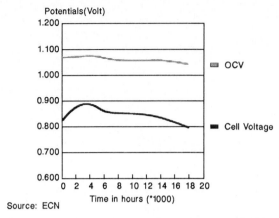

Source: ECN

Fig. 5. Laboratory cell (3 cm^2) performance OCV and cell voltage.

One laboratory cell of 3 cm^2 has now been in operation for nearly 2000 hours. It still produces 814 mV at 480 A (see Fig. 5).

At this moment the accent in the R&D programme changes from aiming at high power densities to improvement of the lifetime behaviour of improved and alternative electrodes and matrices. This will be studied in close combination with effects of raising the pressure in a stack.

The Commission of the European Communities approved a joint proposal of ECN and British Gas to develop a 1 kW internal reforming MCFC stack.

The combination of applied research from several disciplines in combination with developing and testing stacks in one institute has proved to be successful. The old aim to attain an internationally accepted position on MCFC research has been reached. We are hopeful that we can maintain this position by establishing a joint venture which has to define the research and development goals for the near future.

Commercialisation of Fuel Cells

Journal of Power Sources, 29 (1990) 87 - 96   87

DEVELOPMENT AND COMMERCIALIZATION OF ON-SITE FUEL CELL IN JAPAN

NOBORU HASHIMOTO

Advanced Energy Conversion Office, Osaka Gas Co. Ltd., 2-95, 3-chome, Chiyozaki, Nishi-Ku, Osaka 550 (Japan)

Activities in Japan to develop the on-site fuel cell and to bring it into the market have been actively carried forward with the phosphoric acid fuel cell (PAFC) and solid oxide fuel cell (SOFC). To achieve this, potential users, such as gas companies, have been supporting and cooperating with American and Japanese manufacturers of fuel cells.

The PAFC project, aiming for the commercialization of the 50 kW to 200 kW-class on-site units, is currently being carried out. For SOFC, an experimental unit with a capacity of 25 kW will soon be tested.

The development and commercialization of the fuel cell for on-site cogeneration systems can be said to be a pressing and important subject because of its expected contribution to energy conservation and to reducing environmental pollution (Table 1).

On-site phosphoric acid fuel cell

Systems manufactured by the American company, International Fuel Cells Corp. (IFC), have been given field tests involving four PC18 (40 kW) units at Japanese gas companies. Testing of the PC25 (200 kW) preprototype model units has also started in Japan.

In joint development with Japanese manufacturers, about 20 units of 50, 100 and 200 kW trial units are now being fabricated by Mitsubishi Electric, Fuji Electric, and others. These will be field-tested by gas, electric power and petroleum companies.

TABLE 1

NOx content in fuel cell exhaust gas

Fuel cell	NOx (ppm, $O_2 = 5\%$)	Measured date
IFC, PC18, 40 kW PAFC	9	14/1/84
IFC, PC25 preprototype, 200 kW PAFC	< 2	4/7/89
MELCO (Moonlight Project), 200 kW PAFC	6	8/6/89
Fuji Electric, 50 kW PAFC	< 2	22/5/89
Westinghouse Electric, 3 kW SOFC	< 2	21/6/88

0378-7753/90/$3.50

For the commercialization of the on-site PAFCs, three gas companies (Osaka Gas, Tokyo Gas and Toho Gas) plan to introduce a total of 21 IFC PC25 (200 kW) units and monitor operations at customer's sites. The three gas companies and Fuji Electric Co. have entered into a contract to conduct a joint development project of commercial units.

Field test of PC18 (40 kW)

Osaka Gas and Tokyo Gas each purchased one IFC PC18 (40 kW) prototype for testing. Tests were conducted on two PC18 units out of 46 field demonstrations in the GRI Field Test Program.

The two field test units for the GRI program supplied power to the actual load and delivered heat at the test site for three years from 1984 to 1987. They operated for 11 400 and 15 600 h respectively.

In performing these field tests, characteristics of on-site PAFC were profoundly understood, including the confirmation of its remarkable response to load change, allowing it to follow instantaneous ramp rates from 0 to 100% in 0.03 s with a voltage variation as low as 3 V.

PC25 preprototype (200 kW)

Four PC25 preprototype model units have been supplied by IFC for testing in Japan: two units by Tokyo Electric Power, one unit by Osaka Gas and one unit by Nippon Oil. The two units introduced by Tokyo Electric Power are already in operation.

The Osaka Gas unit has been installed at Umeda Center Building (Fig. 1), an office complex near Osaka Station, and will be operated connected to the utility grid. The reject heat is planned to be used for the building's hot water supply. The unit is now being checked for operation. This project will be carried out by Osaka Gas in cooperation with Takenaka Koumuten, one of the largest construction firms in Japan, which designed and constructed the Umeda Center Building. The aim of the demonstration

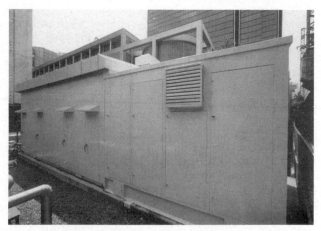

Fig. 1. PC25 preprototype at Umeda Center Building.

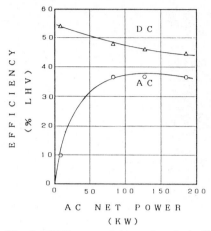

Fig. 2. PC25 preprototype electrical efficiency (%, LHV).

will be to accumulate experience in operation and maintenance of an on-site PAFC. This will serve as a demonstration to future users, as well as to help determine the technical capabilities of this preprototype (Fig. 2).

Commercialization program for the IFC 200 kW unit

IFC is preparing to manufacture the first lot of PC25 units at the present time. They will require an installation area that is smaller than the preprototype model and the unit is expected to have increased performance.

The three major gas companies in Japan, including Osaka Gas, have ordered a total of 21 PC25 units. Although delivery dates are not yet fixed, these will probably be delivered in 1991 or 1992. Most are to be installed at customers' sites and operated for monitoring purposes, in preparation for commercial marketing of PC 25 units.

Development of a 200 kW unit under the Moonlight Project

An on-site PAFC with a capacity of 200 kW (Table 2, Fig. 3) is being developed under the Moonlight Project, with field testing expected this summer (1989) at the Hotel Plaza in downtown Osaka.

The project is being sponsored by NEDO, with Mitsubishi Electric being charged with component development and manufacture of the 200 kW trial unit, and Osaka Gas and Kansai Electric Power carrying out the field tests and evaluating performance.

For this project, a new cell structure was developed and cell characteristics and cell durability were improved. In addition a new compact reformer was developed. Based on these improvements, a 200 kW system was designed and one unit was constructed. In May through July of 1989, component adjustments and power generation tests were conducted at the manufacturer's facilities. Then the unit was installed at the test site at the end of July, and field tests are planned for two years.

Fig. 3. 200 kW PAFC at Hotel Plaza.

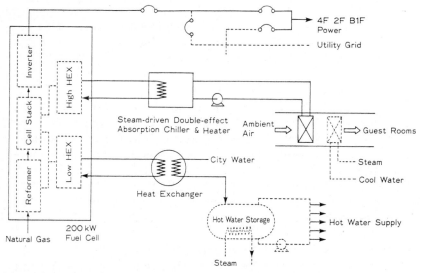

Fig. 4. Moonlight Project 200 kW PAFC field test system.

The unit can be operated both in the grid-connected and grid-independent modes, and will supply hot water to the hotel (Fig. 4). Furthermore, using a double-effect absorption chiller and heater, its rejected heat will also be used to air-condition the building. During the test period, the unit is expected to supply about 10% of both the electric power and thermal demand of this urban hotel, which has 550 guest rooms.

50 kW and 100 kW unit trial manufacture

Tokyo Gas has ordered one 50 kW trial PAFC unit from Fuji Electric and one 100 kW unit from Hitachi. These will be tested at the R&D Institute of Tokyo Gas.

TABLE 2

200 kW PAFC specifications (Moonlight Project)

Rated power (kW)	200
Electrical efficiency (%, HHV)	36 (40%, LHV)
Total efficiency (%, HHV)	80 (89%, LHV)
Fuel	natural gas
Operation	grid-connected, grid-independent
Road following	0 to 100% within 1 min
	20% instantaneous
Heat-up time (h)	3
Total harmonic distortion (%)	less than 2 (grid-connected)
	less than 5 (grid-independent)
Heat recovery temperature (°C)	170 (steam)
	70 (hot water)
Dimensions (m)	W 3.1 × L 10.0 × H 3.2
Cell stack	100 kW × 2
Reformer	100 kW × 2

The 50 kW trial unit has been assembled and continues to undergo power generation tests at the manufacturer's facilities. It will be delivered shortly to Tokyo Gas.

Commercialization of 50 kW and 100 kW unit by three gas companies and Fuji Electric

Three gas companies (Osaka Gas, Tokyo Gas and Toho Gas) and Fuji Electric have jointly begun a new project to develop 50 kW and 100 kW on-site PAFC commercial units.

Based on the technology of Fuji Electric, developed under the Moonlight Project, using the systems design, operation and maintenance know-how accumulated by the three gas companies, this project plans to achieve market entry of commercial units in 1993 (Fig. 5).

In this project, a total of about seventy primary, secondary and production model trial units will be manufactured. Of these 50 production trial

Capacity	Model		1989	1990	1991	1992	1993	1994
50 kW	Trial	Primary		□				
		Secondary			▭			
		Production				▭		
	Commercial						⇐	
100 kW	Trial	Primary				□		
		Secondary					▭	
	Commercial						⇐	

Fig. 5. Commercialization of 50 and 100 kW PAFC.

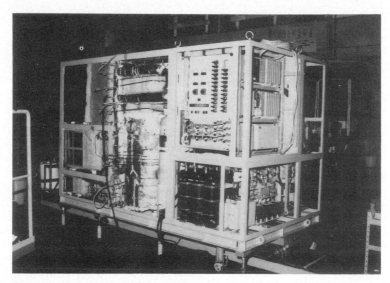

Fig. 6. 50 kW unit for Tokyo Gas.

units will be manufactured in 1991 and 1992. The footprint for the 50 kW package will be 4.0 m^2, its height will be 2.2 m and the target for its electrical efficiency is 40% (Fig. 6).

Other considerations

There are a number of technical and economical aspects which must be addressed prior to the market entry. Among these, the most important is improvement of the cell power density to reduce the package volume and the cost. At this time, power densities of approximately 1 kW/m^2 have been demonstrated using Japanese technology. A cell performance study is actively being conducted to increase power density to 1.5 - 2.0 kW/m^2 maintaining the system power generating efficiency at 40% (LHV).

Development of solid oxide fuel cell

Osaka Gas and Tokyo Gas entered into a contract with Westinghouse Electric in 1986, and have sponsored the development of the SOFC at Westinghouse, which has now reached an exceptionally high technical level. Osaka Gas operated two separate 3 kW experimental units in 1987 and 1988 and evaluated their technical status. Next, tests of a 25 kW generator unit are planned (Fig. 7).

The reason Osaka Gas became involved in the development and commercialization of the SOFC, in addition to the PAFC, is because of several technical advantages:

(1) its power generation efficiency is 10% higher than that of the PAFC

Fuel	Program	82	83	84	85	86	87	88	89	90	91	92
H₂+CO	3 Cell Test(9000Hrs)	▭										
	24 Cell Bundle(2000Hrs)			▯								
	5 kW DOE (400Hrs)				▯							
	400 W TVA						▭					
	3 kWx2 OG, TG						▭					
Natural	3 kW GRI								▭			
Gas	10-20 kW DOE									⊏⋯⋯		
Internal	25 kW KEPCO, OG, TG									⊏⋯		
Reforming	25 kW Cogen. OG, TG											⊏⋯

Fig. 7. Westinghouse SOFC program summary.

(2) its reject heat is at several hundred degrees Celsius

(3) it has a high system simplicity, due to its internal fuel reforming capability

(4) it has a highly stable performance, owing to the use of solid state cell materials

Experimental 3 kW SOFC units [1]

These units were designed to generate 3 kW d.c. at maximum power, and contain 8 bundles of cells, each of which has 18 cell tubes. To evaluate the performance and durability of the cell module (144 tubular cells), operational tests were performed, primarily using H_2 and CO mixtures (Table 3).

TABLE 3

3 kW SOFC specifications

Maximum output power	3 kW d.c.
Fuel	H_2, H_2 + CO, reformed gas
Operating temperature	900 to 100 °C
Generator pressure	0 to 20 inch H_2O
Fuel utilization	55 to 85%
Air stoichiometry	3 to 7

TABLE 4

Operational results of the 3 kW generator system

	Osaka Gas #1	Osaka Gas #2	Tokyo Gas
Operating period	Nov/87 - Mar/88	Mar/88 - Aug/88	Nov/87 - Jun/88
Operating time (h)	3012.5	3683	4882
Availability (%)	97.9	99.1	
Average load (kW)	2.02	2.00	2.00
Start-up (times)	4	2	2
	(including 2 at Westinghouse)		

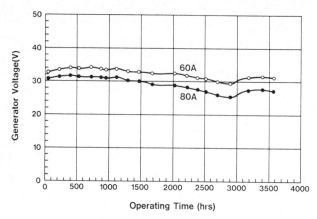

Fig. 8. 3 kW SOFC voltage performance.

Osaka Gas evaluated two cell modules and Tokyo Gas one module, achieving 12 000 hours of total operation (Table 4). The first generator at Osaka Gas was fully tested as planned, but the operation of the second unit at Osaka Gas and that at Tokyo Gas were stopped early due to difficulties with the control system, and not because of cell tube conditions.

It was found that the initial performance of the cell was as expected, but that durability was insufficient. Resistance to thermal cycles was not adequate. After the operation mentioned, the voltage was found to have dropped by 6% at 2 kW output (at about 60 A) (Fig. 8).

In parallel with this project, Westinghouse has improved initial performance, durability and resistance to thermal cycling by improving cell manufacturing methods. The company is, presently, investigating internal reforming methods.

25 kW SOFC generator unit

In view of the technical evaluation of 3 kW experimental units and recent progress in technical improvement by Westinghouse, Kansai Electric Power, Osaka Gas and Tokyo Gas have decided to conduct a joint test of the 25 kW SOFC generator unit equipped with an internal reforming system. This unit will be installed at the research field establishment of Kansai Electric Power, and testing will commence in 1990. The generator consists of two cell modules, each module having 576 cell tubes with 50 cm active length. Its rated power is 25 kW, but can produce a maximum of 40 kW.

Throughout this testing, Westinghouse will accumulate cell production experience and obtain performance data on a large number of cells.

SOFC future project

In addition to the project mentioned above the design and manufacturing of a 25 kW SOFC cogeneration system package is also being studied. This joint development project is being conducted by Osaka Gas and Tokyo Gas along with Westinghouse.

TABLE 5

On-site fuel cell project in Japan[a]

Project		Company	Manufacturer	Power (kW) × units	Operation
PAFC					
Technical development	PC18 Prototype	OG, TG	IFC	40 × 2	1982 - 1985
	PC18 GRI Program	OG, TG	IFC	40 × 2	1984 - 1987
	PC25 Preprototype	OG, TEPCO, Nippon Oil	IFC	200 × 4	1988 -
	Moonlight Project	OG/KEPCO	Mitsubishi	200 × 1	1989 -
	Trial machine	TG	Fuji, Hitachi	50 × 1, 100 × 1	1989 -
	Others	NEDO, PEC, etc.	Fuji, Sanyo, Mitsubishi	50, 100, 200 approx. 20	1990 -
Commercialization	PC25	OG, TG, THG	IFC	200 × 21	(first lot 1991 -) (commercial 1993 -)
	Commercial unit	OG, TG, THG	Fuji	50, 100, approx. 70	trial unit 1990 - commercial 1993 -
SOFC					
3 kW Experimental unit		OG, TG	Westinghouse	3 × 2	1987 - 1988
25 kW Generator		KEPCO/OG/TG	Westinghouse	25 × 1	1990 -
25 kW Cogeneration package		OG, TG	Westinghouse	25 × 2	(1991 -)

[a] OG: Osaka Gas, TG: Tokyo Gas, THG: Toho Gas, PEC: Petroleum Energy Center.

The next step will be field testing of a 100 kW cogeneration system. If technical improvement and cost reduction proceed as expected, it may be possible to introduce the on-site SOFC into the market by 1997.

Economics

Although the cost of the on-site PAFC is 1.0 - 1.5 million yen per kW (7000 - 10 000 dollars/kW), assuming that one unit is ordered at the present time, we believe that through the experiences gained in conducting many PAFC projects, costs for this on-site technology will decrease to the target value of 200 - 300 thousand yen per kW (1400 - 2000 dollars/kW).

Demand for on-site fuel cells will fluctuate sharply depending upon future energy conditions. But, assuming that these remain unchanged from those at present, demand for on-site fuel cell systems in Japan, based on the projected demand for the Osaka Gas service territory, is estimated to be 100 000 kW or more yearly in the mid 1990s, equivalent to 1000 units of 100 kW each, or more. However, in order to disseminate on-site fuel cells in the market, several models of different capacity up to 1000 kW must be developed.

Conclusions

Japan's three major gas companies have been actively promoting the development and commercialization of on-site fuel cells in Japan [2]. Through these activities the know-how for various techniques for systems design, operation and maintenance of on-site fuel cell units has been acquired.

Based on this experience, we will continue to support technical development and commercialization of fuel cell manufacturers and expect to bring PAFC into the market from around 1993, with SOFC to be introduced several years later (Table 5).

References

1 M. Harada and Y. Mori, Osaka Gas test of 3 kW SOFC generator system, *Abstr. 1988 Fuel Cell Seminar, Long Beach, CA, U.S.A., Oct. 1988.*
2 N. Hashimoto, Overview of the Osaka gas on-site fuel cell program, *Ext. Abstr. Int. Seminar Fuel Cell Technology and Applications, The Netherlands, Oct. 1987.*

Journal of Power Sources, 29 (1990) 97 - 107

TOKYO ELECTRIC POWER COMPANY APPROACH TO FUEL CELL POWER PRODUCTION

TOYOYASU ASADA and YUTAKA USAMI*

*Engineering Development Center, The Tokyo Electric Power Co., Inc.,
1 - 3 Uchisaiwai-Cho 1 chome, Chiyoda-Ku, Tokyo 100 (Japan)*

Introduction

Fuel cells, which were first described in the 19th century as a direct power generation method, are currently regarded as being viable. Energetic efforts are continuing mainly in the U.S.A., Japan and Europe to establish fuel cell power as a new technology for residential and commercial sectors. Fuel cells can be classified into the alkali type and the phosphoric acid type, which belong to the 1st generation; the molten carbonate type and the solid oxide type which belong to the next generation. Different components and system configurations are used due to the differences in electrolytes and operating temperatures. From the standpoint of use, they are classified roughly into fuel cells for on-site generation (several hundred kilowatts) and those for distributed generation (several megawatts).

The following describes the Tokyo Electric Power Company's involvement in the development of phosphoric acid, molten carbonate and solid oxide fuel cells. It introduces the company's plan to develop phosphoric acid fuel cells which are considered the most practical at present. An 11 MW plant under construction in Chiba Prefecture is outlined and the themes and trends of the development are reviewed.

Development of fuel cells

Development of phosphoric acid fuel cells

During the latter half of the 1970s, the Tokyo Electric Power Company recognized the importance of research and development (R&D) for distributed power sources located near consumers as well as concentrated large-capacity power sources for stabilizing power supply in the future. As fuel cells are a most promising new power source for this purpose, TEPCO began investigation and research.

In 1980, we introduced a 4500 kW phosphoric acid water-cooling experimental plant (supplied by United Technologies Corporation) from the

---

*Author to whom correspondence should be addressed.

0378-7753/90/$3.50

U.S.A., which represented the most advanced technology at the time in this field. The tests were started at the Goi Thermal Power Plant outside of Tokyo, Japan. In April 1983, we succeeded in generating power and continued the tests until December, 1985. Since then, many technological ideas have been acquired and a foothold in the technology for practical use has been established. Table 1 shows the generation performance at the end of the test period.

After thoroughly studying the results of the tests, it was decided to introduce an 11 MW plant from Toshiba (fuel cells were manufactured by International Fuel Cells (IFC)). TEPCO made a significant contribution to the design by conducting a feasibility study, making various improvements in component development. Foundation work began in January 1989, with plans to commence testing in January 1991.

For independent power supplies, TEPCO began to develop a standard atmospheric pressure air-cooled 200 kW-class system in 1986 using Japanese technology. Testing began at the Shin-Tokyo Thermal Power Plant in September, 1987. Subsequently, two 200 kW class systems were imported from IFC in 1988. One was installed and tested in the Shin-Tokyo Thermal Power Plant with the air-cooled system. The other was installed and tested in a building in Shibaura, Tokyo.

Development of next generation type fuel cells

A feasibility study for a molten carbonate fuel cell to be used in a power plant was begun in 1984. Based on the feasibility study results, the

TABLE 1

Operational records of a 4500 kW plant

(1)	Power generation test	
	Test period	April 83 ~ Dec. 85
	Initial power generation	April 83
	Rated power generation	Feb. 84
(2)	Cumulative gross electric power	5428240 kW h
(3)	Cumulative operation practice	
	Power generation	2423 h
	Stand by	464 h
(4)	Maximum continuous power generation	500 h
(5)	Hot times	
	CSA	4233 h
	Reformer	4098 h
	Auxiliary burner	3787 h
	TMS burner	1661 h
(6)	Cycles	
	CSA	50
	Reformer	68

power plant showed great promise, particularly in view of its compatibility with a coal gasification plant. TEPCO then began fundamental research in 1986. Presently, this research is being directed to problems concerning various materials of the fuel cell and its technological characteristics.

Separate investigations and research on a solid oxide fuel cell were begun in 1985. Presently, components are being developed by making a prototype concurrently with investigation and research.

The Tokyo Electric Power Company considers that the next generation of systems can be put to practical use only after development of a phosphoric acid system is realized. That is, use of molten carbonate and solid oxide fuel cells for practical purposes is not possible without sufficient development and experience of a phosphoric acid fuel cell.

Commercialization of phosphoric acid fuel cells

Development policy

Recently, industries, government agencies and universities have shown an increasing interest in cogeneration systems due to their effective energy utilization and cost reduction.

In fact, cogeneration installations using gas turbines, gas engines, diesel engines, etc., have markedly increased recently. The Tokyo Electric Power Company believes that fuel cells will become important cogeneration systems in the future due to their efficiency and environmental advantages. This belief is based on test experiments of a 4500 kW plant and a 200 kW air-cooled system.

Fuel cells are classified into an independent supply type and a regional supply type. On the basis of market survey results, a 200 kW class plant and a 10 MW class plant are adequate in size for independent (on-site) supply and regional (distributed) supply, respectively. These two projects will now be discussed.

Development of 200 kW system for individual supply

A system for independent supply was installed in hotels, hospitals and other buildings which have a large demand for heat. Presently, the 200 kW air-cooled system, which was developed jointly with the Sanyo Electric Co., and the 200 kW water-cooled system, which was purchased from IFC, are being tested. The air-cooled system and one of the water-cooled systems are being tested in the Shin-Tokyo Thermal Power Plant. The other water-cooled system was installed in the basement of the Kandenko Building located in Shibaura in Tokyo. Plans have been made to investigate the characteristics and reliability of 200 kW class fuel cell power plants and study their suitability for independent power supply. Table 2 shows the specifications of the air-cooled and the water-cooled systems, and Scheme 1 shows the test schedule. Figures 1 and 2 show flow diagrams for these systems, respectively. The following results were obtained from tests of the 200 kW water-cooled fuel cells.

TABLE 2

Specifications of a 200 kW test plant

	Air-cooled (N200)	Water-cooled (PCX)
Rated power (kW (a.c.), gross)	220	200
Output voltage (V (a.c.))	210	210
Minimum power (kW (a.c.), gross)	50	50
Controlled power range (%)	22.7 - 100	25 - 100
Electrical efficiency (% gross, HHV)	35	35 ~ 38
Waste heat recovery rate (%)	40	45
Overall thermal efficiency (%)	75	80 ~ 83
Base fuel	city gas	city gas
Fuel consumption at rated power (Nm^3/h)	60	45 ~ 48
Start-up time (cold, h)	4	5
NOx emission (ppm)	⩽ 30	⩽ 25
Noise level at site boundary (dB)	⩽ 50	⩽ 50

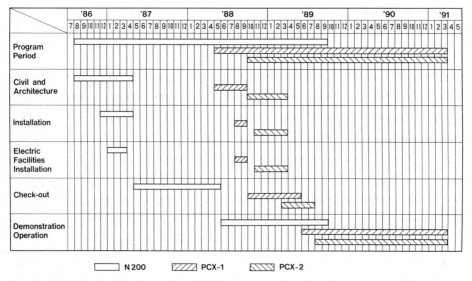

Scheme 1. Test schedule of a 200 kW plant.

(1) Rated generation was achieved (200 kW).

(2) Generation efficiency nearly matched the design efficiency, namely, 35% (transmission end, HHV).

(3) Start up time up to initial load was 3.5 h while the designed value was 5 h (cold start).

(4) Speed of output change was 15 s for 25% - 100%, the target value.

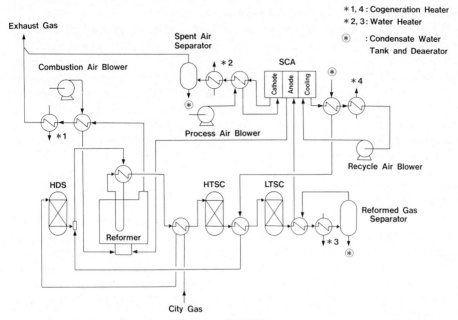

Fig. 1. System flow of an air-cooled type 200 kW plant.

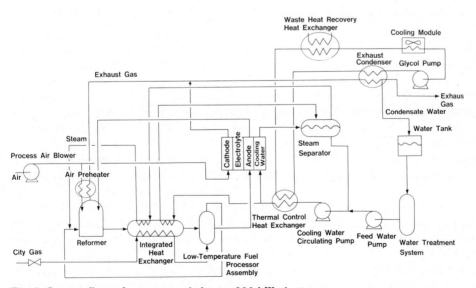

Fig. 2. System flow of a water-cooled type 200 kW plant.

(5) Harmonic voltage distortion was about 1.2% during the rated operation while the designed value was below 3% (overall).

(6) Environmental characteristics were extremely good. NOx, SOx and dust levels were below the detection limits at the exhaust tower outlet.

Development of 11 MW system for regional supply

TEPCO began construction of an 11 MW test plant at the Goi Thermal Power Plant in order to develop a regional supply system which could be constructed in urban redevelopment sites and other areas having a large energy demand. This is the world's largest fuel cell plant and was developed jointly with the Toshiba Corporation and the IFC Corporation, on the basis of the 4500 kW experimental plant test results. The Tokyo Electric Power Company hopes to obtain information regarding development of a commercial system through this experiment. The plan of the 11 MW experiment plant is described below.

(1) Test items

The test aim at verifying the suitability of a phosphoric acid fuel cell as a distributed urban power source by studying the following items: (i) demonstration of a power and heat supply system, (ii) verification of plant operation performance, (iii) major component verification performance, including the fuel cells and reformer, (iv) establishment of reliable operation technology targeting automatic operation, (v) confirmation of favorable environmental characteristics.

(2) Brief description of plant

This plant is based on PC23, which was developed by the IFC Corporation and the Toshiba Corporation as a standard system for the American market. We adopted new improvements in the major components, such as the cells, reformer, turbo compressor and the inverter, on the basis of accumulated experience from the 4500 kW plant. Efforts were made to attain high reliability by testing the plant components separately at full scale in advance and by making improvements where problems occurred. The specifications were reviewed and partly changed so as to ensure compatibility with site conditions in Japan. Tables 3 and 4 show the planned performances and the specification of the principal machinery, respectively; Fig. 3 shows the system flow sheet and main characteristics. The efficiency and the capacity of cell stacks were increased by utilizing fuel cells having a large area, higher temperature and pressure operation than the earlier 4.5 MW plant. The heat load conditions of the reformer were made less severe and the NOx level was decreased by introducing exhaust gas combustion and adopting a bottom combustion burner. Circuits of the inverter were simplified by installing a large-capacity GTO. The quality and the quantity of recovered heat and the recovery cost were analyzed. Three locations which can recover heat economically were selected. Data on the range of effective utilization of waste heat and operation characteristics will be obtained.

As the adjacent area has no large demand for heat, it is planned to discharge the recovered heat by exchanging it to sea water in the future.

TABLE 3

Planned performance of an 11 MW plant

Rated power at 66 kV utility grid	11 MW (a.c.)
Output voltage	66 kV
Minimum power	0 MW
Controlled power range	30 - 100%
Controlled VAR range	—11 to +11 MVA
Electrical efficiency based on fuel HHV at utility grid	41.1%
Waste heat recovery rate based on fuel HHV	31.6%
Overall thermal efficiency (HHV)	72.7%
Base fuel	liquefied natural gas
Fuel consumption at rated power	2100 Nm^3/h
Start-up time (cold/hot)	6/2.5 h
NOx emission	$\leqslant 10$ ppm
Noise level at site boundary	$\leqslant 55$ dB (A)

TABLE 4

Specifications of 11 MW principal machinery

Machinery	Item	Specifications
(1) Cell stack assemby	cell stack arrangement	6(series) × 3(parallel)
	unit cell dimension	100 cm × 100 cm
Full size stack	unit cell output	670 kW
2000 h operation	operating pressure	7.4 kg/cm^2 g
	operating temperature	207 °C
(2) Reformer	reformer tube material, number	superalloys, 54
Full size simulator	operating pressure	approx. 10 kg/cm^2 g
Burner characteristics	combustion method	spent air mixing
Temperature distribution		lower portion burner
(3) Turbocompressor	type	generalized design
		lateral, 2 shafts
Full size model	arrangement	2 stage series with
		intercooler
Start-up operation characteristics	lubrication	oil
	start-up	air injection to compressor blade
Burner characteristics	delivery pressure	approx. 8 kg/cm^2 g
(4) Inverter	capacity	11800 kV A
Full size inverter, reactor	bridge number	3 pair × 18 poles
Transformer, controller	power device, number device	GTO thyristor, 36
Operation characteristics	device rated value	5000 V × 2500 A
	control signal transmission	optical fiber
	exciting method	self exciting
(5) Waste heat recovery	waste heat recovery rate	
	(HHV, gross)	41.3%
	(HHV, net)	31.6%
	heat recovery facility	CSA cooling water cooler
		fuel gas cooler
		cathode spent air cooler

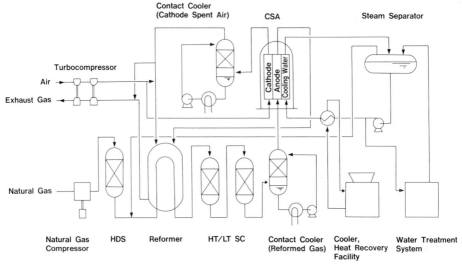

Fig. 3. System flow of an 11 MW plant.

	'88	'89	'90	'91	'92	'93
	8 9 10 11 12	1 2 3 4 5 6 7 8 9 10 11 12	1 2 3 4 5 6 7 8 9 10 11 12	1 2 3 4 5 6 7 8 9 10 11 12	1 2 3 4 5 6 7 8 9 10 11 12	1 2 3 4 5
Program Period						
Pallet Fabricaiton						
Civil and Architecture						
Installation						
Electric Facilities Installation						
Check-out						
Demonstration Operation						

Scheme 2. Test schedule of an 11 MW plant.

(3) Schedule (Scheme 2)

The Tokyo Electric Power Company began this construction project in July 1988, through site preparation, with the civil engineering work starting on January 20th, 1989. Toshiba's Keihin Plant began production of pallets (a pallet is a unit which has built-in machinery process vessels and pipes) in October 1988. Pallets produced at the plant will be site-installed and connected with pipes to complete the plant. Construction work for machinery and pallet installation will be carried out throughout the year. PAC (Process and Control) tests will be started in February 1990. An approximate 2-year test period will begin in January 1991.

Commercialization obstacles

In order to commercialize a new technology, it must be both technologically and economically sound. A fuel cell has the following characteristics: (i) improvement of operation reliability, (ii) compactness of system (footprint reduction), (iii) reduction of construction costs. The inherent characteristics of a fuel cell for regional supply will be discussed later with reference to the 11 MW plant, which is now under construction.

Operation reliability
The reliability of the cell itself must be raised. In small laboratory tests using simulated gas, the cell exceeded 40 000 h in some cases. According to field tests, the record of a GRI plan (40 kW) is 15 000 h and that of a MW class system is about 2500 h.

Cell life must be at least 40 000 h and preferably more than 60 000 h for purposes of commercial use.

On the other hand, the service life of the reformer is another factor in achieving high reliability. The need for compactness and rapid load following is required to offset the severe increase of the reformer's heat load change. This has been the principal cuase of plant problems.

Compactness
The 11 MW plant now being constructed is about 0.3 m²/kW, and is larger than that of present competitive technologies. The footprint should be as small as possible to meet the requirements of cogeneration systems being planned for urban areas. This is an important factor, particularly in Japan, where land prices are high. Our target for commercialization is 0.1 m²/kW in the case of a 10 MW class plant.

Construction cost
In the case of the 10 MW plant, the current construction cost is approximately 1 million yen/kW, which is still higher than that of existing systems. It should be lowered to 200 - 300 thousand yen/kW (in consideration of the merits of the fuel cell). The merits of fuel cells include lower NOx levels, greater thermal efficiency, lower transmission and distribution costs, and reduced transmission loss due to the vicinity of the consumers. Some reports on the future forecast of fuel cell costs indicate that their competitive strength in costs can be increased sufficiently by the establishment of a mass production factory. The largest reduction cost factor

is the fuel cell modules which currently account for 50% of the total plant costs. It is predicted that the manufactured cost of the cells can be lowered to almost 1/10 by technological improvements and mass production.

Measures for solving problems and future prospects

Greater efforts must be made to solve the problems discussed in the preceding section. The Tokyo Electric Power Company is considering the following steps for attaining these targets.

Simplification of system

The performance and reliability of individual machinery and the plant by testing the 11 MW plant will need to be confirmed. The system will be simplified by eliminating the sensors which are incorporated for diagnostic purposes. Other simplification measures include eliminating redundant machinery, reducing the pallet size and simplifying the system on the assumption of attaining fully automatic/unmanned operation. Plant space can be reduced by adopting a hierarchical structure, by installing a reformer and cells in the basement level and by rationally reducing maintenance space.

Promotion of technological development

In parallel with the tests, development of the following component technologies will be promoted to attain greater compactness and to reduce costs. For example, TEPCO will increase the capacity of the stack and improve its durability by increasing the cell's power density and developing a large area cell. Also, performance of the reformer will be raised by improving catalyst performance, heat transfer method, etc. The inverter bridges and d.c. circuits will be improved and other improvements made on electrical machinery. The low temperature carbon monoxide reactor will be eliminated, efficiency of the turbo compressor will be raised and failure diagnosis technologies will be developed.

Operation of semi-commercial plant

TEPCO plans to design an improved plant on the basis of the afore-mentioned improvements, and will confirm that these technologies are suitable for the purposes of commercial use. TEPCO will also establish operational technologies, including the applicability to a power heat supply system, and will construct and operate the plant according to its design conditions.

Efforts for mass production

The last step is mass production. At this stage, the construction costs can be lowered considerably by a mechanization/automation process, mastery/standardization of operations, and by cost reduction of materials and machinery through bulk purchase. In other words, the achievement of mass production is the key to commercialization. Naturally, fuel cell demand must increase to enable mass production. This is difficult to achieve by the efforts of only one company or one country. International cooperation is essential. Learning from the past PC 23 Project, the American Public Power Association (APPA) recently announced a NOMO (Notice of Market Opportunity), which requests cell makers to promote commercialization of phosphoric acid fuel cells for distributed installation. Cooperation between makers and users as proposed in this report is essential. In addition to these technological efforts, administrative cooperation is also necessary. For example, the related legal regulations must be improved, tax credits must be given and long-term low interest loans must be provided.

Conclusions

The development of fuel cell power generation, which is expected to become a distributed power source in the future by using phosphoric acid fuel cells, has been discussed. There are many problems still to be worked out for commercialization of fuel cells. These include the development of individual components, such as cells and reformers, the optimization and tests of an integrated system, the improvement of manufacturing technology for cost reduction, and the promotion of mass production by increasing fuel cell demand. A multi-lateral approach is necessary in order to overcome these difficulties and to successfully realize phosphoric acid fuel cell power generation. There is also an urgent need to grasp the current situation accurately and to device an overall strategy. Cooperation among different industries and countries is a prerequisite.

Journal of Power Sources, 29 (1990) 109 - 117

FUJI ELECTRIC PHOSPHORIC ACID FUEL CELL ACTIVITIES

RIOJI ANAHARA

Fuji Electric Co., Ltd., 12-1, Yurakucho 1-chome, Chiyoda-ku, Tokyo 100 (Japan)

Introduction

Fuji Electric Company (FE) has been engaged in the research and development of fuel cells since 1961. In recent years, we have concentrated on the development and manufacture of phosphoric acid fuel cell (PAFC) targeting mass-production and full commercialization in 1994.

FE's PAFC development can generally be divided into two categories ranging from kilowatt to megawatt range:

(i) stationary PAFC for dispersed power plants, on-site, and cogeneration applications

(ii) transportable PAFC for vehicular and remote site applications

Figure 1 shows a summary of the application of our PAFC plants. Other fuel cell technologies under development at FE are alkaline fuel cell, molten carbonate fuel cell and solid oxide fuel cell.

We are confident that we can help to meet the challenge of the future via the successful development and commercialization of PAFC for both stationary and vehicular applications, and that we can help provide a highly efficient alternative, thus contributing to a cleaner global environment on the Earth.

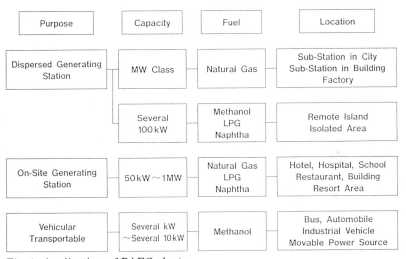

Fig. 1. Application of PAFC plants.

0378-7753/90/$3.50

PAFC for stationary, cogeneration applications

Fuji Electric has accumulated since the late 1970s construction and operation experience of PAFC plants from kW to MW ranges. Now total projected and expected PAFC plants of FE have reached more than 70 projects, 12 000 kW under patronage from both domestic and overseas customers (U.S.A., Europe and South East Asia).

1000 kW PAFC plant (NEDO Project)

The plant was constructed as a dispersed power station for electric utility purposes in the premises of Sakaiko Thermal Power Station of Kansai Electric Power Company, Japan and has been successful as expected. The total accumulated generation hours exceeded 2000 h inclusive of 500 h of non-interrupted operation. The main operation results and recommended performance improvements are shown in Tables 1 and 2, respectively.

The average voltage deterioration rate, about which we are most concerned, is shown in Fig. 2. Real performance data was gained only up to 2000 h but the extrapolated curve to 40 000 h shows the average deterioration will be 2.5 micro V/h. This may be reasonable at the current technology level but it is not satisfactory for the future commercial stage.

TABLE 1

Operation results of 1000 kW PAFC plant (NEDO Project)

Item	Target	Results
Electrical output	1000 kW	1000 kW
Electrical efficiency (HHV)	40%	37.1% (40.87 achievable)
Generating hours	> 1000 h	2045 h (including 705 h non-interrupted)
Reformer operation hours		3614 h
Cold start-up time	4 h	5 h
Load response	25 - 100%/min	±11%/min (125 - 450 kW)
Shutdown time	1 h: normal	47 min: normal
	⟨1 min: emergency⟩	10 s: emergency
Environment	NOx < 20 ppm	NOx 10 ppm
	SOx < 0.1 ppm	SOx 0 ppm

TABLE 2

Performance improvements (1000 kW PAFC plant, NEDO Project)[a]

1. More stable and longer operation required
2. Cell performance should be improved
3. Uniform temperature distribution in reformer is necessary (shorter start-up time, quicker load response)
4. Higher efficiency expected

[a]From NEDO's report.

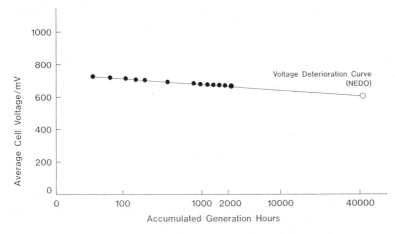

Fig. 2. Average voltage deterioration rate (1000 kW PAFC plant).

After completing the demonstration operation, the plant was dismantled and key components such as the fuel cell stack, reformer, turbo compressor etc. were examined in great detail.

Based on the design and operational experiences of the 1000 kW plant, NEDO and electric utility companies in Japan are planning to construct larger capacity dispersed PAFC power stations with the cooperation of manufacturers.

200 kW PAFC plant for a remote island (NEDO Project)

The purpose of this plant is to demonstrate the feasibility of a PAFC plant as the power supply system on a remote island. The plant has been operated at full power in our factory and will be shipped to the site (Tokashiki Island, one of the most southern islands in Japan, Okinawa Electric Power Company) in the middle of September 1989. The plant is expected to begin operation at the site from the beginning of October 1989.

To meet the special conditions in this remote island, the plant uses methanol as its primary fuel and operates in parallel with diesel generators.

Table 3 shows the key specification of the plant and Fig. 3 shows a view of the plant under construction.

50 kW PAFC plant for electric utility companies

We have been operating a 50 kW pressurized PAFC plant for Tohoku Electric Power Company (one of our electric companies which supplies electric power to the northern area of Japan) since March 1987 with good performance. LNG is used as the primary fuel for the first phase operation and LPG in the second phase.

The total operation time exceeded 8000 h as of the end of August 1989 (including 1000 hours unmanned non-interrupted operation and also including 590 h operation with LPG), which is the longest operation record

TABLE 3

Key specification of PAFC plant for remote island (NEDO Project)

Purpose	generating station in a remote island
Type	water-cooled PAFC
Output	200 kW (sending end)
Efficiency	>37% (HHV) (sending end)
Construction	packaged type (outdoor use)
Fuel	methanol
Working pressure	atmosphere
Minimum load	20%
Start-up time	<3 h (cold start)
Load response time	30 s for 20 - 100% load, instantaneous for ±20%
Voltage distortion	<3% total
Operation mode	independent and/or grid connected, self-start, self-synchronized
Weight	<20 t (as package)
Footprint	<0.2 m²/kW (as package)

Fig. 3. General view of 200 kW PAFC plant for remote island (under construction) (NEDO Project).

of pressurized PAFC systems in the world. The success of LPG reforming should be noted also as the first experience in Japan.

Table 4 shows the key specifications and operation record of the plant.

50 kW PAFC for gas utility companies

The 50 kW plant for the Tokyo Gas Company is now being tested, in FE's facility very satisfactorily, and will begin operation at the site from the beginning of October 1989.

TABLE 4

Key specification and operation record of 50 kW PAFC plant (for Tohoku Electric Power Company)

Type	pressurized, water-cooled PAFC
Output	50 kW (sending end)
Fuel	natural gas, LPG
Working pressure	2.0 kg/cm^2 A
Working temperature	190 °C
Cell area	3600 cm^2 class
Hydrogen utilization factor	80%
Air utilization factor	55%
Operation hours	8000 h (at end of Aug. 1989) (world's longest operation of a pressurized PAFC plant)
Fuel	LNG/LPG (first LPG operation in Japan)
Grid connection	first connected operation with distribution networks in Japan
Exhausted NOx	6 ppm

TABLE 5

Key specification of 50 kW PAFC on-site plant

Purpose	on-site co-generation
Type	water-cooled PAFC
Output	50 kW (sending end)
Electrical efficiency	40% (LHV) (sending end)
Total efficiency	80% (LHV)
Construction	packaged type (indoor/outdoor)
Fuel	town gas (LNG) (13 A)
Working pressure	atmosphere
Range of output	0 ~ 100%
Start-up time	<4 h
Voltage distortion	<5% total
Operation mode	independent and/or grid connected
NOx	<10 ppm (O$_2$: 7%)
Weight	<5 ton
Footprint	<0.11 m^2/kW (L: 3.1 m, W: 1.75 m, H: 2.2 m)

This plant is very compact and will serve as the precursor for future commercialization. Grid connected as well as independent power operations are possible. The key specification and outer view of this on-site plant are shown in Table 5 and Fig. 4 respectively.

Recently three major gas utility companies in Japan jointly requested Fuji to supply nine 50 kW plants before 1991 and seven 100 kW plants by 1993. The Japanese Government also intends, beginning in 1990, to demonstrate 10 to 16 sets of 50 kW PAFC plants, grid-connected in conjunction with solar and wind power systems.

114

Fig. 4. Outer view of 50 kW PAFC plant for cogeneration.

TABLE 6

50 kW on-site PAFC plant-order list

Item	Customers	Quantity	Delivery
1	Tokyo Gas	1	1989
2	NEDO, Kansai Electric Power, Rokko Island	> 10	1990 - 1992
3	EGAT (Thailand) ODA Project[a]	1	1991
4	Europe	1	1990
5	Tokyo Gas	4	1990 - 1991
6	Osaka Gas	4	1990 - 1991
7	Toho Gas	1	1991

Natural gas	Ambient pressure	Water cooling

[a]ODA: Official Development Assistance.

Table 6 shows the list of 50 kW PAFC on-site plants for which we have received orders.

Fuji is now offering the current type 50 kW plants to several potential customers in Europe, U.S.A. and also South East Asia markets.

General purpose small capacity PAFC

Many of our customers have expressed interest in using fuel cells in various new applications. We have already supplied several 4 kW methanol reforming, normal pressure and air-cooled portable fuel cell units. It is very difficult for such small capacity PAFC plants to be economically competitive with other conventional systems unless some new innovative technologies are applied.

PAFC for vehicular applications

In 1983 Fuji Electric and Engelhard Corporation, which had developed a substantial amount of expertise in PAFC technology, entered into a co-operative agreement to develop PAFC power packs for forklift truck application. Several prototypes were developed and tested. Currently, Fuji Electric has full responsibility for commercializing fuel cell powered forklifts since Engelhard's management decided to terminate their fuel cell program.

The development and engineering of the Fuji Electric PAFC power pack for forklift trucks are essentially complete. In order to penetrate into the potentially large market for such trucks in the U.S.A. and Europe we need to substantially reduce the production cost of our product. Our current forklift program goal is just that: cost reduction.

It is our belief that the initial commercialization of PAFC product(s) for vehicular applications, where the unit size of the fuel cell required is larger, and/or where there is a definite need for the injection of a new technology, can succeed.

For bus application where a 50 kW unit is called for (*versus* 5 kW for forklift truck), we feel that the cost goal can be met. In addition, the environmental regulations mandated for 1991 and 1994 in the U.S. offer the opportunity to introduce a 'clean' technology.

Finally, since Fuji's PAFC is methanol-fed, it will help to reduce the country's dependency on foreign oil in the future. We at Fuji Electric believe these forces provide a strong impetus to PAFC development and commercialization.

U.S. Department of Energy (DOE) bus project

Fuji Electric has committed several million dollars towards the development of a forklift PAFC system. The basic technology and experience gained in our forklift development has served as the basic design of the DOE bus power system. This is a very compact, liquid-cooled stack design, fuel cell–battery hybrid propulsion system, which has been developed and demonstrated through the operation of the forklift power pack; it could be applied to the PAFC bus system without any major modification. Therefore, minimal research and development are required to convert the forklift design to buses. The merit of the liquid cooling technology is summarized in Fig. 5. The probability of success is quite high.

The DOE bus project is composed of 4 phases, that is,
Phase 1 (1987 - 89): proof of feasibility for fuel cell/battery subsystem
Phase 2 (1989 - 91): proof of concept for fuel cell/battery subsystem
Phase 3 (1991 - 92): test-bed bus evaluation
Phase 4 (1992 -   ): prototype bus testing

For phase 1, the half size 25 kW brassboard including chopper has been shipped to Chrysler Pentastar Electronics, which is responsible for the fuel cell–battery hybrid propulsion system for the bus, after having satisfactory

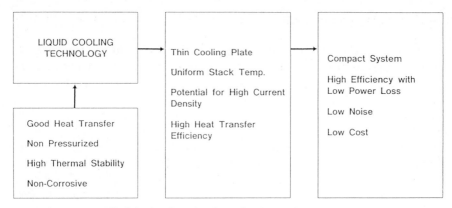

Fig. 5. Features of Fuji fuel cell technology for fuel cell bus system.

TABLE 7

Key specifications for 50 kW fuel cell system

Items	Phase I	Phase II
Configuration	brassboard	onboard
Net output	22.5 kW	50 kW
System efficiency	36% (LHV) (excluding auxiliary power)	38% (LHV) (including auxiliary power)
Chopper		
Efficiency	90%	95%
Frequency	1.3 kHz	10 kHz
Auxiliary power	utility power	battery power
Cold start-up time	45 min	25 min
Standby to full power (0% to 100%)	7 min	3 min
Interface	analog	digital

approved tests at Fuji's Chiba Factory. The plant has begun combined testing with the peaking battery at Chrysler since the middle of July 1989.

Table 7 shows the comparison of the 25 kW brassboard plant and the 50 kW full size.

The overall combined tests of fuel cell and battery hybrid system at Chrysler Pentastar Electronics will last until the end of December 1989 and we are expecting successfully to enter into phase 2 of the program where 3 sets of 50 kW PAFC plants will be manufactured and tested.

Conclusions

Our commercialization plan for stationary and vehicular PAFC has been highlighted. Our goal is to make PAFC a commercial reality, with appropriate cost and benefit trade-offs to users. We believe that environmental and energy concerns will be critical to PAFC's success in the coming years, and that a most attractive market for PAFC in the future lies in the U.S., especially in Southern California, Japan and European countries.

Collaborative efforts between foreign organizations and Fuji are necessary to promote and demonstrate fuel cell applications, as efficient and environmentally superior alternatives, for a cleaner Earth.

Journal of Power Sources, 29 (1990) 119 - 130

THE FUTURE OF FUEL CELLS FOR POWER PRODUCTION

JEFFREY A. SERFASS

Technology Transition Corporation, 1101 Connecticut Avenue NW,
Washington DC 20036-4303 (U.S.A.)

The fuel cell, which William Robert Grove conceived of in 1839, has engendered high hopes over the last two decades. Conventional technologies have served power production needs well over the course of the 20th century, but those with vision have begun to anticipate a time when a new generation of technologies would emerge — technologies that are close to environmentally benign, with characteristics more of solid-state devices than of rotating machines, with high efficiency, in small packages. Some have linked technologies to be used after the year 2000 with a fuel that is also environmentally benign — hydrogen.

Developers of generating equipment see the opportunity to move the manufacturing and construction operations from the field to the factory, producing truck-transportable pallets of equipment. Electric utilities see a path to a reduced burden from the environmental impact of their smoke stacks and by-product wastes and from the growing network of transmission lines. Researchers see an opportunity to participate in the development of the first completely new technology for electric generation since nuclear power. And, environmental advocates see the opportunity to support a technology capable of having a major impact on global atmospheric and water problems. At least, it is hoped these interest groups have these visions.

In many respects it seems that fuel cells have been the elusive solution in search of a problem, or problems, yet to develop. Their environmental performance begs for a day when air and water quality constraints are so great that fuel cells are the only option one has for power production. Their fuel flexibility characteristics beg for a day when fuel availability and price considerations require a technology that can utilize low-Btu waste gases from landfills, and natural gas, and methanol with great efficiency and flexibility. These and other scenarios have kept public and private funding of fuel cells somewhat steady since the 1970s.

The capital and operating cost of the fuel cell as a power producer has received attention, but any disadvantageous comparison has been augmented with fuel cell 'credits' that are based upon one or more constraints on existing technologies.

This orientation has put fuel cell developers in the position of chasing constraints that continue to be 'in the future', not 'today' constraints. Air quality problems, in general, have only begun to suggest pervasive problems

   © Elsevier Sequoia/Printed in The Netherlands

in building power plants. In the meantime, progress in developing fuel cells has still not reached the goals of cost and reliability necessary for entering the commercial market. The *potential* for the product and the *potential* market have yet to merge into a viable commercialization setting.

Long-term opportunity — is it still there?

In examining the future of fuel cells for power production, look first at the long-term need for power production capability and technologies. Are fuel cells still a promising option?

Environmental considerations

In the last decade, the U.S. electric utility industry has spent staggering amounts on pollution control: capital investments for SO_2 emissions control alone were more than $60 billion. The annual costs for air pollution control exceed $10 billion; for solid waste disposal, $1 billion. And these amounts, of course, do not include the additional hundreds of billions of dollars spent to improve nuclear plant safety or in the cancellation of planned nuclear capacity or for other environmental expenditures for non-generation operations.

The summer of 1988 with its unprecedented temperatures — reinforcing a pattern through the 1980s — again focused major attention on the environment. Concerns over global warming are reinforcing action to address other areas of environmental concern, including acid rain and ozone depletion.

With the call for increased attention to the environment by both national parties, environmental constraints facing the deployment of new electrical generation will only grow. These concerns will likely involve emphases on higher efficiencies, cleaner fuels and wise energy use.

Substitution of fuel cells for conventional power plants should improve air quality and reduce water consumption and waste water discharge. The generation of electricity now produces more particulates, sulfur oxides and nitrogen oxides than all other stationary sources combined. Fuel cell power plant emissions are ten times lower than those specified by the most stringent environmental regulations. Fuel cells also produce lower carbon dioxide (CO_2) emissions than conventional generation, a question of increasing concern due to the so-called 'greenhouse effect'.

Because the electrochemical reaction of the fuel cell produces water as a by-product, little if any external water is required for power plant operation. This low water use is in marked contrast to large steam electric power plants that require large quantities of water for cooling. Waste water discharges from fuel cell systems are also lower and the quality is superior compared with conventional fossil-fueled power plants, scarcely requiring any pretreatment prior to disposal in many communities. Fuel cells eliminate or reduce water quality problems associated with thermal discharges, power plant site runoff, and the disposal of wastes from air emission controls.

The quiet, electrochemical nature of fuel cells eliminates many of the sources of noise associated with conventional steam-powered systems, thus easily complying with OSHA (Occupational Health and Safety Administration) standards. No ash or large volume wastes are produced from fuel cell operation. Land requirements are acceptable, and connecting transmission corridors are not required as is the case with outside power sources. Because of their comparatively small size, absence of a combustion cycle, state-of-the-art safety systems, and low pollutant emissions, fuel cells are among the least hazardous methods of energy conversion.

Size considerations

Planning flexibility, including modularity, results in strategic and financial benefits to the utility and its customers. Because fuel cell power plants may be built within two years from the time of order, and because performance is largely independent of plant size, they can be used to increase utility system capacity by small increments in response to customer needs.

By better matching increases in electric demand, long periods of over-capacity are avoided, lowering average fixed costs over time. And, if demand growth is uncertain, the fuel cell's short lead time becomes even more valuable. A utility can slow or accelerate its response to growth. Also, as experience is gained with fuel cells, utilities may be able to reduce required reserve margins while maintaining the same reliability, resulting in lower fixed costs.

Efficiency considerations

The fuel cell can convert up to 80% of the energy from its supply fuel into useable electric power and heat. Current phosphoric acid fuel cell (PAFC) designs offer 41% electrical conversion efficiency on a high-heating value basis, with 46% electrical conversion efficiencies for PAFC possible in the near term through currently known science and engineering. The Electric Power Research Institute has estimated that advanced molten carbonate fuel cells (MCFC) may achieve electric efficiencies greater than 60%, exclusive of a bottoming cycle, which could raise efficiencies even higher. Such efficiencies are unprecedented. Furthermore, a fuel cell's efficiency is largely independent of its size. Fuel cells can operate at half their rated capacity while maintaining high fuel-use efficiencies.

Fuel cell power stations located close to loads can also reduce costly transmission lines and transmission losses. Possibly more important, fuel cells sited in municipal systems can minimize transmission line dependence in joint power supply arrangements.

Another important attribute of the fuel cell is its ability to cogenerate; that is, to produce hot water and lower-temperature steam at the same time as it generates electricity. Its ratio of electric to thermal output is approximately 1.0, while for a gas turbine the ratio is about 0.5. This advantage means that a fuel cell matched to a thermal load will have approximately twice the electric output of a combustion turbine matched to that same load*. In smaller sizes of interest to most public power systems, fuel cells are also more efficient (by about a factor of two) when compared to, for example, the 15 000 Btu/kW h heat rate of a 2 MW combustion turbine. The fuel cell's load following capability, while maintaining high efficiency, may also give it an advantage in cogeneration markets with varying heat demands.

Fuel considerations

Though initial fuel cell power plants may be designed to be fueled primarily with natural gas, the cells require hydrogen. The fuel processor that produces this hydrogen-rich gas allows the use of a variety of low-sulfur gaseous and liquid fuels including propane, methanol and ethanol. Advanced fuel cells should also be able to operate economically on these fuels as well as gasified coal.

Operational considerations

Fuel cells have beneficial operating characteristics matched by no other technology. These characteristics save costs in meeting system operating requirements. Dynamic operating benefits include load following, power factor correction, quick response to generating unit outages, and control of distribution line voltage and power quality control.

The solid state power conditioning system of the fuel cell power station can be used to control real and reactive power independently. Control of power factor and line voltage, to meet load can minimize transmission losses and reduce requirements for reserve capacity and auxiliary electrical equipment such as capacitors, tap-changing transformers and voltage regulators.

When new generating capacity is added to an electric power system, substation equipment sometimes has to be upgraded because of the expectation of increased fault current (which lowers the reliability of the electric system). However, with fuel cell power units, it is not necessary to upgrade the fault-current interrupting capability of existing substation equipment, because the short circuit generated will be limited.

Fuel cell units have an excellent part-load heat rate and can respond rapidly to transient loads. For example, the heat rate of a phosphoric acid demonstration unit is anticipated to be approximately 8300 Btu/kW h at rated power and to increase only slightly at 50% of rated capacity. Also, it is expected to be able to ramp from 30% of rated power to 100% of rated power in only 7 s. Spinning reserve requirements can thus be lessened when fuel cells are used.

Cost and design considerations

The fuel cell power plant, as a long-term option, must produce electricity at costs competitive with today's alternatives. Without this expecta-

---

*Dependent upon a user's thermal load requirements it may be economical to install boiler systems to supplement the cogenerated heat produced by fuel cell systems.

tion for a generally competitive technology, the manufacturers and the early buyers will not absorb the higher costs and risks of the market entry activities.

The cost target identified by several electric utilities considering 11 MW demonstrations in 1987, was approximately $1100/kW (1989 $). A recent market study [1] identified a similar figure for public power systems, for a long-term mature cost goal.

Phosphoric acid technology is expected to yield power plants that are only marginally competitive in a setting with low coal prices, and abundant gas and oil supplies, especially when new technology risk factors are considered. The installed cost of molten carbonate and solid oxide technologies is expected to be favorably competitive by a margin of 10% or greater.

The expectation that a technology can achieve competitive costs is important to establishing the basis for proceeding with commercialization. Site-specific applications that yield high 'credits' for the absence of certain constraints can provide high value opportunities for limited-production early-market units, but they are not useful for evaluating the mature market.

This expectation of a competitive technology is more easily realized if the basis for design is simple, with few integrated thermal loops, limited rotating machinery, and minimum maintenance due to water treatment, waste-product removal and catalyst regeneration.

In summary, the longer-term power production business is likely to require a technology with the characteristics of fuel cells. Power producers will require these characteristics in smaller-size packages suitable for modular capacity expansion, and small enough to serve individual community requirements. The technology and the power plant design must be simple enough that the early buyer can expect that cost reduction efforts will be successful. And, the power produced must have costs competitive with power produced from alternative means.

Fuel cells fit these long-term needs, expected to develop beyond the year 2000. Cost considerations lead one to favor the advanced fuel cell technologies, but phosphoric acid, with progress made toward simplicity of design and low stack costs, is also a contender.

Near-term opportunity — how do we reach the long term?

In the near term, the circumstances favorable for fuel cell commercialization require identification and development of the right technologies, aggressive manufacturers, and a market that needs the characteristics described above *today*. Even with this, the additional support of government policy (national and local) and research funding may be necessary. Without the combination of factors described above (technology, manufacturer, market), and barring severe environmental constraints, getting through the costly, risky early-market phase will be difficult.

In the U.S., one sector of the electric utility industry is trying hard to find the right technology/manufacturer combination to work with as the early market. Public power systems are looking for a commercial-scale demonstration and market entry program that leads directly to commercial products in the mid-1990s.

The American Public Power Association (APPA) initiative, described by their October 1988 Notice of Market Opportunity [2], identifies the interest of municipally-owned electric utilities. These potential fuel cell buyers need the characteristics of fuel cells, today, in relatively small packages. Their systems are generally smaller in size, serving urban loads, with siting, environmental and transmission constraints. Some municipal systems have difficulty today siting any conventional technology within their service territory.

Profile of public systems

Public power represents 14% of the U.S. electric utility industry measured in kilowatt hour sales to ultimate customers, and 12.5% of the industry in terms of installed generating capacity. The public power sector is made up of approximately 2000 individual municipal systems, 60 joint action agencies which supply groups of member systems, and other wholesale suppliers (see Scheme 1).

Some salient characteristics of this market are:
● Public power is growing at a faster rate than the industry as a whole (4% *versus* 2.1% growth in kW h sales to ultimate customers)
● Public power retail rates are often lower than surrounding private and rural cooperative utilities (an average of 35% lower than the industry as a whole)
● 83% of the public power energy sales to ultimate customers is supplied from purchased power contracts
● The largest 20 municipal systems represent about 37% of the retail energy (kW h) sales of public power systems
● Three-quarters of the systems are above 10 MW in size, and about 150 systems have peak loads in excess of 100 MW
● Public power has 89 000 MW of installed capacity today
● Approximately 18% of the energy purchased by public power systems is supplied by joint action agencies; about 900 systems are members of joint action agencies

Public power structure

The public power segment of the U.S. electric utility industry is made up of individual systems that sell power within the approximate boundaries of cities, small and large, and so-called public utility districts (located principally in Nebraska and the Pacific Northwest). The largest system is owned by the City of Los Angeles with a load of 4750 MW and the smallest have peak loads less than one MW.

MWH SALES
(Ultimate Customers)

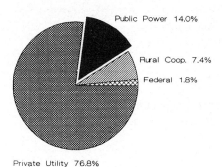

Public Power 14.0%

Rural Coop. 7.4%

Federal 1.8%

Private Utility 76.8%

NO. OF SYSTEMS
(Include Non—Retail Systems)

Private Utility 6.2%
213

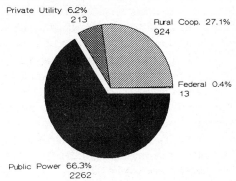

Rural Coop. 27.1%
924

Federal 0.4%
13

Public Power 66.3%
2262

Scheme 1.

Power supply for the smaller systems is usually purchased from a neighboring private or cooperative utility on a rate schedule that may include both a firm contract for power, to which both the seller and buyer are committed, and 'partial requirements' of varying power supply which may fluctuate as load and other factors vary. Many public power systems generate some of their own power, for instance, with peaking turbines or diesel engines to reduce the higher cost peak load purchases.

The municipal utility is managed by a utility staff, with oversight by an elected city council, or an elected or appointed governing board. As entities of a political subdivision, the management of a public power utility is likely to be influenced by citizen, political and technical considerations, as well as economic factors. Many public power systems have responsibility for more than electric sales, including gas, water, sewage and even cable TV.

Joint action agencies, such as the Municipal Electric Authority of Georgia, have been created in many regions of the country to allow smaller public power systems to meet power supply needs collectively. A joint action agency might represent 5 to 60 systems, in a single state, and have a peak load of several hundred megawatts or more. They are large enough to build their own generating capacity, and many do. They usually do not own the transmission system that connects their widely dispersed membership. As organizations formed and supported by individual member utilities, joint action agencies are more likely to be influenced by economic and risk management factors and interest in self-dependence than public acceptance and political factors. Oversight is by a 'utility-knowledgeable' board.

Public power decision-making

A workshop for APPA was conducted by George Mason University, Center for Interactive Management, in Fairfax, VA, to explore factors that would impact a fuel cell market penetration strategy. The two-day session involved invited public power guests with an interest in their own electricity generation. The objectives were to identify and set priorities among factors influencing decisions regarding self-generation and to identify barriers inhibiting the purchase of self-generation technologies.

Several expected utility factors emerged as important:
- Financial and cost considerations
- Impact of rates to retail customers
- Reliability

But, several decision factors, possibly more important to public power systems than other utilities, also emerged:
- Political and public acceptance
- Environmental concerns
- Site availability
- Control of own destiny

The traditional factors for evaluating generation additions — *cost and risk* — are important to public power. Experience in working with the utilities participating in APPA's recent fuel cell demonstration project development efforts, however, suggested two important characteristics of these factors in the public power market.

(1) Certain public power systems may be willing to pay more for fuel cells than traditional economic comparisons would suggest. They value the environmental and operating features quite highly, including the relatively small size offered by fuel cells.

(2) The risk that many smaller public power systems are able to accept in deploying a new technology like fuel cells is relatively low. A fuel cell power plant of 10 to 50 MW may comprise such a large part of their capacity requirements that its performance must be highly reliable, even for early units.

Public power needs, or at least desires, new generation owned by municipal or joint action agency organizations. The deregulated direction that the U.S. electric utility industry is taking may provide a more competitive utility business, on the one hand, and more difficult inter-utility business transactions on the other. This trend, plus the traditions of public power, argue for replacing power purchases with self-generation if the costs and other factors tilt decision-makers in that direction.

Fuel cell features of interest

The interest of municipal electric systems in the fuel cell evolves from its numerous attractive, and unique combination of, features.

Size. At one to 50 MW, no other technology offers such high efficiency. This technology is sized to meet the smaller capacity requirements of public power.

Part load efficiency with fast response. Many systems desire a technology capable of economic operation in a broad, intermediate duty range, possibly to include baseload operation under certain circumstances.

Cogeneration potential. Many municipal systems will prefer cogeneration installations to increase public acceptance and the initial economics. Fuel cells are easier to site than other cogeneration technologies.

Environmental attractiveness. Urban power systems with generation within their boundaries will increasingly require very clean technologies. The lack of significant water requirements by fuel cells is also a real plus in a significant portion of the U.S.

Low noise. City-sited power plants must have noise characteristics that make them unobtrusive neighbors.

Short lead time and modularity. This adds up to building the capacity very close to the time that you need it. This is a real strategic benefit.

The reasons for utility and joint action agency interest offer significant insights that lead to the methodology for assessing the market.
   • Certain utility characteristics seem to favor fuel cells, but the reasons for being an advocate or a leader vary significantly from utility to utility.
   • Some municipal systems would prefer to own both peaking and intermediate supply capability, and they expect to continue to purchase baseload capability, often because of its low cost. Others, however, *do* foresee a baseload dispatch mode for fuel cells, especially with cogeneration.
   • The desire for self-generation is rooted in interest in controlling one's future, particularly controlling costs.
   • Systems that view themselves as environmentally constrained, today or eventually, are the most likely potential buyers of fuel cells. Some of these systems believe it is the *only* technology that they could build on or near their system.
   • The first buyers will be manager-advocates with a vision of the future into which this technology fits. The boards or city councils will share at least the key elements of their vision.
   • Joint action agencies appreciate that this technology has the size and other characteristics such that it can be sited at individual member systems. Therefore, in addition to its other advantages, fuel cells may reduce the need for transmission capacity linking their cities.
   The potential of fuel cells to break through the efficiency barriers now being met by all conventional generation cannot be overemphasized. As a developing technology, fuel cells should see significant efficiency improvements. Over the next 15 to 20 years, more advanced fuel cell designs such as molten carbonate and solid oxide may be able to demonstrate efficiency improvements in the order of 50% or greater.
   The fuel cell's competitors, on the other hand, including gas turbines and internal combustion engines, are at a mature stage of development. Small incremental improvements are the most that can be expected from these technologies. And improvements in conventional technologies will

come at the expense of higher operating temperatures and, therefore, greater nitrogen oxide air pollution.

The market potential in public power is large

The quantification of the market focuses on three applications:

(1) additions to meet load growth requirements

(2) replacement of retired generation

(3) replacement of purchased power contracts

It is this third factor that is relatively unique to public power. It is the independence from purchased power that most strongly drives public power interest in fuel cells.

The period of interest is from 1996, the earliest date when fuel cells may reasonably be expected to be commercially available, to 2010, a date by which more advanced technologies may be considered.

The absolute market potential appears to be about 89 000 MW in the 1996 - 2010 study period. The estimated maximum market potential, or technical market, utilizing conservative values for key parameters (load growth and maximum dependence on a single technology), is 37 000 to 44 000 MW over the 15-year period. Nearly 40% of this market is outside of the largest 20 utilities. Although, many public power systems are small, 93% of the public power market can be met with a unit size no smaller than 10 MW.

Of the 37 000 to 44 000 MW maximum market potential, 14 000 to 17 000 MW is considered to be the conservative estimate of the 'potential early market.' These are utilities

(1) Whose rates are lower than the neighboring large utility, thereby giving some room for purchase of a higher-cost supply resource

(2) That are located in an air quality or water availability constrained area

An additional screening, to isolate utilities having an interest in self-generation or having existing generation, yields the 'likely early market' of 28 000 to 30 000 MW, utilizing base case values for key parameters. The more conservative parameter values yields a 12 000 to 14 000 MW market. Even this lower market estimate would provide an ample base for early market development.

Other factors may also play a role in the early fuel cell market within municipal systems.

● Transmission constraints faced by public power encourage development of self-generation over purchases.

● Interest in using the power plant as a cogenerator may have a positive or negative impact on fuel cell commercialization, depending upon whether the utility or thermal-user owns it.

● The risk of early units — technical, financial and strategic risk — must be born to some extent by the buyers and the seller. Public power has indicated a willingness to accept some of this risk, but very small systems may not be able to.

• Natural gas pricing and availability is not expected to adversely impact fuel cell commercialization, although long-term gas contracts or partnerships with the gas supplier may overcome any concerns that do exist.

• Global warming and other environmental concerns have heightened overall sensitivity to environmental issues and may have an impact on local issues.

Screening curve analysis of phosphoric acid and molten carbonate fuel cells, in competition with purchased power and the generating technologies that are available to smaller systems, indicate that a capital cost below $1100/kW (1989 $) will be necessary to achieve a sizeable mature market. Phosphoric acid fuel cells may not reach a price this low, but this is within the range of expectation for molten carbonate fuel cells. Solid oxide fuel cells also offer the promise of low capital cost and high efficiency but were not considered in this analysis.

The path to a mature product at a mature cost begins with a high-cost product. Competition with alternatives, and numerous factors affecting the early market determine the size of the expanding production base and the magnitude of declining product costs. Utilities constrained by environmental factors or size were examined separately to allow for the more limited competition that the fuel cell will see in these segments.

The results indicate that phosphoric acid fuel cells can achieve a mature market of almost 1200 MW/year at a cost of about $1100/kW. Public power is not a sufficient market to drive the costs lower. Therefore, phosphoric acid fuel cells may have difficulty competing in unconstrained public power systems. The addition of other potential markets — private utilities and others — could help to lower capital costs and thereby increase the market share of phosphoric acid fuel cells within public power.

Molten carbonate fuel cells appear more attractive, achieving a market of almost 2000 MW/year at a cost of $925/kW. Public power is sufficient for a production volume that might yield competitive costs, but even molten carbonate fuel cells will be unable to compete well with combined-cycle in unconstrained markets without a much larger market from other utilities or users.

Of course, conventional alternatives to fuel cells may, themselves, face more stringent environmental performance standards. As the costs of power from conventional alternatives may climb, the cost-competitive thresholds for fuel cells would also rise. This change in competitive position would lead to either higher capital cost thresholds for fuel cells to achieve the market levels noted above, or would lead to larger markets at the original thresholds.

Forward pricing of fuel cells, where the initial price of the product is lower than initial production costs, may be necessary to establish the market share that fuel cells are capable of attracting.

Public power presents a significant mature market for fuel cells but the real opportunity is in matching the unique needs of many public power systems to the early market needs of the manufacturers. Benefits to both public power and to manufacturers will accrue if both proceed to develop this opportunity.

Conclusions

Public power systems provide only 15% of the electricity sold in the U.S., but they can provide a market for fuel cells of 1000 to 2000 MW per year over the first 15 years of fuel cell commercialization. With their current need for a technology like fuel cells, and with the knowledge that they can provide a significant early market, APPA has asked interested fuel cell developers to present a viable demonstration and market entry program for their consideration. The opportunity presented is one of a collaborative effort to push through the early market obstacles.

The recent market study [1] for the Electric Power Research Institute and APPA concludes that a technology likely to yield a competitive product ($1100/kW 1989 $), promoted by a supplier capable of supporting product guarantees in the early commercialization stages, will have an eager market in public power systems.

The future of fuel cells for power production, therefore, is bright. Technology is ready for commercial-scale demonstrations. At least some of the fuel cell technologies offer the promise of electricity produced at competitive costs. The future will require the characteristics of fuel cells for power production. And today, in Europe, Asia and North America, there are constrained areas that can provide the market for early commercialization steps.

The long-term opportunity exists. The short-term need exists in certain market niches. In the U.S. it is in the public power sector of the utility industry. What remains to be developed is one or more collaborative buyer/ seller commercialization programs that push through the near-term obstacles.

References

1 Technology Transition Corporation, *The Market for Fuel Cell Power Plants Within Municipally-Owned Electric Utilities*, Oct. 1989.
2 American Public Power Association, *Notice of Market Opportunity for Fuel Cells*, Oct. 1988.

European Fuel Cell Interests

Journal of Power Sources, 29 (1990) 133 - 142

FUEL CELLS IN EUROPE

P. ZEGERS

Commission of European Communities, rue de la Loi 200, B-1049 Brussels (Belgium)

Introduction

Ten to fifteen years ago fuel cell research and development (R&D) in Europe flourished with extensive R&D on phosphoric acid (PAFC), molten carbonate (MCFC), solid oxide (SOFC) and alkaline (AFC) fuel cells.

Around 1975, most of these activities had stopped with the exception of AFC. The year 1985 was a turning point where R&D on most fuel cell types started again in fuel cell R&D programmes in the Commission of European Communities (CEC), F.R.G., Italy, the Netherlands, Spain, Norway and Switzerland. Since 1985 much has happened:

• A 1 MW PAFC plant will be operational in 1991 in Milan, Italy
• A 1 kW ER-MCFC became operational in July 1989 in the Netherlands
• Promising results have been obtained in the development of direct methanol fuel cells (DMFC) for small-scale application and transportation
• Three projects with a strong industrial participation will start to develop a 10 kW MCFC plant with coal gas and two 1 kW SOFC units, fueled with methane.

Before discussing European fuel cell research in more detail, some general aspects of fuel cells are discussed.

Fuel cells as a major technology for energy conversion

At present, energy conversion is 90% based on combustion of coal, oil and gas, and combustion processes still have a long way to go. The efficiency of these processes however is often not very high (Carnot), their pollution is considerable and the problem of CO_2 is becoming increasingly important.

A change to other more efficient and less polluting systems should therefore be seriously considered. Electrochemical energy production (fuel cells) and storage systems (secondary batteries) are an interesting possibility.

These systems could cover many areas.

• Large (MW) scale power production with efficiencies up to 70% (instead of 40% obtained with steam turbines); these fuel cells may also be used in seaships (PAFC, MCFC, SOFC).
• Cogeneration systems which produce power with 50% efficiency and in addition industrial process heat at 600 or 900 °C (MCFC, SOFC).

0378-7753/90/$3.50

TABLE 1

Specification of primary energy use in Europe (1030 million tonne oil equivalent)

Energy use	Proportion of total (%)	Fuel cell types
Buildings	43	
Heating	25.8	PAFC (80 - 120 °C)
Electricity	17.2	
Industry	37	PAFC
Electricity	15.5	MCFC (650 °C)
Process heating	21.5	SOFC (1000 °C)
Transport	20	PAFC
		DMFC (70 - 100 °C)
		SPFC (70 °C)
Seaships	±18	PAFC
		MCFC
		SOFC

● Cogeneration systems which produce electricity at 40 - 45% efficiency and heat at 80 - 120 °C for heating of buildings (PAFC).

● In transportation, electrical vehicles with fuel cells may achieve efficiencies which are 2 to 3 times higher than petrol engines (DMFC, PAFC, AFC).

From Table 1, which gives a specification of the primary energy use in Europe, it is clear that cost effective fuel cells can play a major role in all energy demand sectors.

Some data on fuel cells and factors influencing their application

The major components of a fuel cell power plant are given in Fig. 1. The fuel processor (reformer for methane or naphtha and coal gasifier for coal) transforms fuel into hydrogen gas which in its turn is converted into electricity and water in the fuel cell. A large variety of systems is possible, depending on the type of fuel cell and on how the waste heat of a fuel cell will be used: cogeneration, internal reforming, bottoming cycle. Much work will have to be done on the optimization of a fuel cell plant for particular applications. Some data on the most important fuel cell types are given in Table 2.

Contrary to combustion engines, fuel cells have high partload efficiencies (for some fuel cells such as PAFC and DMFC partload efficiencies are even higher than the fuel load efficiency). Generally, a rapid power variation is possible for MW size plants (10 → 100% in 30 s). Finally, the fuel utilization in the fuel cell is 80 - 85% and the remaining 15 - 20% H_2 will have to be used in an efficient way.

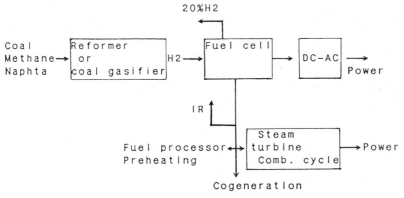

Fig. 1. Schematic layout of a fuel cell plant.

TABLE 2

Fuel cell types and properties

Fuel cell power plant	Efficiency methane → electricity (compl. plant) (%)	Temperature waste heat (°C)	State of art
AFC	35 - 40	60 - 80	10 - 100 kW
PAFC	35 - 42 (47)	80 - 120	1 - 4 MW
MCFC	65 - 70	500 - 600	20 kW
SOFC	65 - 70	700 - 900	5 kW
DMFC	methanol → power 40 - 50	60 - 100	200 W

Internal reforming

An important concept for fuel cells fueled with methane and operating at temperatures higher than 600 °C (MCFC and SOFC) is the concept of internal reforming (IR). Here, waste heat of the fuel cell is used to transform methane into hydrogen *in* the fuel cell. In this way, an external reformer is not needed which can lead to a 30% cost reduction. In addition, the cooling requirements are lower which leads to an additional cost reduction. The efficiencies for IR fuel cells are generally somewhat higher than fuel cells which use external reformers.

Problems with internal reforming mainly arise for MCFC which require reformer catalysts due to the fact that the waste heat is available at 600 °C and the reforming temperature for methane is around 800 °C. Much R&D is still needed to find suitable catalysts which can resist the very corrosive environment of MCFC. SOFC which operate at 1000 °C give less problems.

Modularity

Modularity is another important topic. A modular system obviously has advantages:

● The size of the installation can be adapted to the demand and this leads to reduction in capital cost

● Dispersed installation (e.g. in towns) is possible due to the low pollution of fuel cells and their modularity; this leads to lower power distribution costs

The extent of modularity however depends very much on the type of fuel cell power plant. In a coal gasifier with a fuel cell and a combined cycle, both the coal gasifier and the combined cycle have a strong economy of scale whereas only the fuel cell (with only 20% of the total cost) is modular. On the other hand, a methane fueled internal reforming fuel cell with only a small part of the electricity delivered by a steam turbine has a strong modular character.

A methane fueled fuel cell therefore can be designed for both modular and less modular applications. A coal fueled fuel cell plant generally has a rather strong economy of scale.

Pollution

Pollution in fuel cells is generally an order of magnitude smaller than for combustion systems. Of the two main components the fuel cell itself contributes very little to pollution; the main source of pollution is therefore the reformer or the coal gasifier.

Methane reforming occurs at 800 °C for which heat is delivered by a combustor which produces most of the NOx. With proper design of the burner NOx pollution can be brought down to 5 ppm. In the case where an internal reforming is used a combustor is not needed and the NOx pollution can be reduced to 1 ppm. SO_2 pollution is generally negligible and the hydrocarbons amount to 3 - 30 ppm. Figure 2 gives a comparison of the pollution of conventional and fuel cell power plants.

When *coal* is used, the fuel processor is a coal gasifier which transforms coal into a hydrogen rich gas. Also in this case the coal gasifier is the main cause of pollution. This pollution however is considerably lower than coal combustion of powdered coal. In fact, several countries consider replacing powdered coal combustion by coal gasifiers (with a combined cycle) for environmental reasons. If coal gasifiers will be used in future fuel cells may become very attractive from the point of view of cost, pollution abatement (a combined cycle produces NOx and a fuel does not) and energy saving. Data for NOx and SO_2 in gram per GJ electricity produced, are given in Table 3. The low value of SO_2 for MCFC plants is due to the fact that the sulfur tolerance for MCFC is 1 ppm and the gas from the coal gasifier has to be desulfurized to that level. The hydrocarbon content of exhaust gases in coal fueled fuel cell plants is negligible.

A rough cost estimate of NOx, SO_2 and CO_2 extraction is given in Table 4.

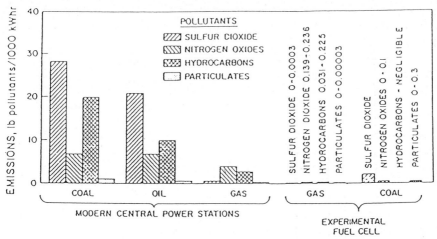

Fig. 2. Comparison of power system emission.

TABLE 3

NOx and SO$_2$ pollution in coal fueled power plants

	Coal combustions	Coal gasification	Coal gas + PAFC	Coal gas MCFC	Coal gas SOFC
NOx (g/GJ)	150	20	1.6	1.6	1.6
SO$_2$	1000	24	24 (24 ppm)	1 (1 ppm)	24 (24 ppm)

TABLE 4

Cost of CO$_2$, NOx and SO$_2$ extraction

	Coal combustion	Gas combustion	Coal gasification
Cost CO$_2$ extraction	5 ct/kW h[a]	3	2
Cost NOx + SO$_2$ extraction	2 - 4 ct/kW h		

[a]Dutch currency 235 ct = 1 ECU.

Finally, the pollution from fuel cells used in *transportation* can be expected to be very low. Fuel cells such as AFC, using hydrogen as a fuel, have no pollution at all. Methanol fueled fuel cells where methanol is directly oxidized or is transformed into hydrogen by internal reforming, have extremely low pollution levels. Methanol fueled PAFC with an external methanol reformer can also be expected to have the highest pollution. Comparison of a methanol PAFC system and an internal combustion engine is given in Fig. 3.

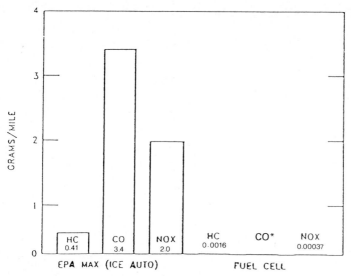

Fig. 3. Comparison of pollution levels for ICE and fuel cell driven cars (*CO not detect-able with equipment sensitivity of 100 ppm).

Fuel cell R&D in Europe

An overview of European fuel cell R&D is given in Table 5. In Europe about 21 MECU/year is spent on terrestrial applications and 9 MECU/year for space applications.

European fuel cell R&D is generally following the lines given below.
● Europe should carry out R&D on all major fuel cell types for large-scale power production (PAFC, MCFC, SOFC). MCFC and SOFC seem more power production (PAFC, MCFC, SOFC). MCFC and SOFC seem more attractive due to higher efficiencies and higher waste heat temperatures. The technical barriers may however turn out to be insuperable and PAFC may increase its efficiency in small steps to 50%. At present, therefore one cannot say that there is a winning concept and R&D on all three fuel cell types is needed.
● CEC and national programmes should try to be complementary.
● Collaboration between all European fuel cell programmes should be promoted.

PAFC

The situation in Europe on PAFC may be characterized as follows:
● Europe has no PAFC stack producers
● PAFC know-how is available in Europe with companies such as Johnson Matthey and AEG
● Europe is strong in reformers (Haldor Topsoe, KTI) and a.c.–d.c. conversion

TABLE 5

European fuel cell programmes

	Start of programme	Duration (years)	Budget (MECU)	Fuel cell types
CEC	1989	3	25 (CEC 50%)	SOFC MCFC DMFC PAFC SPFC
Netherlands	1986	5	30	MCFC PAFC
Italy	1986	5 (a new 5 year programme is being prepared)	40	PAFC MCFC
F.R.G.		2 MECU/year		SOFC AFC
Spain	1988	5	15	MCFC
Norway	1988	5	3.5	SOFC
Switzerland	1988	3	4.5	SOFC
ESA	1988	2	18	AFC

This situation lead to a concept where European companies design and construct PAFC plants which have Japanese or U.S. PAFC stacks but where all other components are delivered by European manufacturers.

A major project is the construction of a 1 MW PAFC pilot plant in Milan, Italy by Ansaldo and Haldor Topsoc with stacks from the U.S. company IFC. This project is funded by the Italian FC programme with some financial support from the CEC.

Four 25 kW PAFC pilot units are being constructed by KTI with Italian, Dutch and CEC funding; the stacks will be delivered by Fuji.

MCFC

Dutch, Italian and CEC programmes on MCFC started in 1986 (Spain started in 1989). Before that date know-how on MCFC in Europe was very small.

In the Netherlands, ECN started a technology transfer programme with I.G.T. from the U.S., which was very dynamic and well structured. This lead to the operation of a 1 kW ER-MCFC unit in July 1989. In the future, 2.5 and 10 kW pilot units are planned in 1990 and 1991 respectively. The participation of an industrial partner is a key to the continuation of this programme. In the period from 1986 to 1991 30 MECU will be spent on Dutch MCFC research.

In Italy, R&D is following the same lines with stack development up to 10 kW MCFC units during the next three years and basic R&D on new cathode materials, nickel dissolution, corrosion problems, etc... ENEA is responsible for the overall management of this programme in which ANSALDO, CISE, CNR and a number of universities participate. For the period 1986 - 1989 around 5 MECU has been spent.

CEC, MCFC R&D started in 1986. During the first three years, research was focussed on basic R&D. An important task of the CEC here was to promote collaboration between European fuel cell programmes in particular in the field of basic R&D. In 1989, a new three-year programme was defined which includes:

- Basic R&D
- Development of a 1 kW internal reforming MCFC stack
- Development of a 10 kW MCFC pilot plant for coal gas

The total cost of this programme is 8 MECU of which the CEC contributes 40%.

Spain has started a 13 MECU five-year MCFC programme this year.

SOFC

At present SOFC work in Europe is carried out in the CEC, F.R.G., Norway and Switzerland. In Italy, a programme is being prepared. Know-how in Europe is comparable to that in the U.S.A. and Japan due to:

- Large SOFC programmes in the past (e.g. BBC, F.R.G.)
- Extensive work on high temperature electrolyzers which are closely related to SOFC (Dornier, F.R.G.)

In Europe industrial interest in SOFC is strong. This is possibly due to the fact that SOFC offers good possibilities by combining SOFC with existing power production technologies such as steam turbines or combined cycles.

The CEC started a two-year exploratory SOFC programme in 1987. Basic research was carried out to develop new electrode and electrolyte materials. Five SOFC concepts (2 honey combs and 3 flat plates) were also investigated. Finally, a market and a system study were carried out.

The market study by GEC was focussed on two applications:

- 200 kWe units for industrial cogeneration
- 200 MWe SOFC + combined cycle plants for power production

For a three-year period (1997 - 2000) a market of 50 - 100 MWe is predicted for the 200 kWe cogeneration units. Until 2015 a total SOFC market of 80 000 MW is expected. Most promising markets are F.R.G., U.K., Italy, Spain and the Netherlands. The expected efficiency and cost is given in Table 6.

The system study carried out by TNO investigated a 200 kWe cogeneration unit and a 100 MWe power plant consisting of a SOFC with a steam turbine bottoming cycle. The study investigated different SOFC options: ER or IR SOFC and different systems where the contribution of the steam

TABLE 6

Efficiency and cost of SOFC plants

	Efficiency (%)	Cost with present state-of-the-art (ECU/kW)
200 kWe CHP	80 (electricity + heat)	900 - 1000
200 MWe	70 electricity	800 - 900 (including combined cycle)

turbine bottoming cycle in the total electricity production varied from 12% to 40%.

The studies lead to the conclusion that cost effective SOFC can be developed with existing materials. A SOFC development plan was developed with the following targets:

- 2 × 1 kW SOFC units in 1992
- 1 or 2 × 20 kW SOFC in 1995
- 200 kWe cogeneration unit in 1997

The ongoing CEC SOFC programme (1989 - 1992) includes the following topics:

- Construction of a 1 kW SOFC stack with flat plate cells and metal bipolar plates (Siemens)
- Construction of a 1 kW SOFC with a modified tube concept (ABB)
- Basic R&D on electrodes and electrolytes
- Construction of a 100 W flat plate SOFC unit with a ceramic bipolar plate (Imperial College)

The total budget for three years is 12 MECU of which 50% is paid by the CEC.

SOFC R&D is also carried out in F.R.G. (tube type SOFC), Norway (thin-film electrolyte SOFC) and Switzerland.

Fuel cells for small-scale power plants and transportation

Fuel cell applications for transportation have the advantage that they can lead to efficiencies which are two to three times better than petrol engines and that the pollution is more than one order of magnitude smaller. The problem however is that the cost is around 500 - 1000 $/kW which, for a 40 kW car, is far too expensive. FC driven electrical cars however may have a chance due to:

- More severe environmental regulations (e.g. no ICE cars in the center of towns such as Rome and Milan)
- R&D aiming at a strong cost reduction of fuel cells

The CEC programme is focussed on cost reduction and has the following objectives:

- Development of methanol fueled fuel cell concepts where the methanol reformer can be deleted; this could lead to a 50% cost reduction and a much less bulky fuel cell plant
- Increase the current density, to reduce cost per kW
- Development of a fuel cell concept which allows cheap mass production
- Reduce the amount of precious metal catalyst

From 1985 to 1988 CEC research was investigating two ways to delete the methanol reformer:

- Development of a direct methanol fuel cell (DMFC)
- Development of methanol internal reforming fuel cell (IR-SPFC) operating at 300 °C.

This research lead to the development of a new catalyst which allowed operation of DMFC for 4000 h without poisoning of the catalyst.

R&D on IR SPFC, focussed on development of suitable solid electrolytes operating at 300 °C, was less successful. Several electrolytes have been developed but they could not be used at temperatures higher than 150 °C.

Future CEC research 1989 - 1992 will continue work on DMFC; the work on IR SPFC has been stopped. The objective is to develop DMFC with solid electrolytes operating at 100 to 150 °C with 100 - 150 mA/cm^2, 0.6 V and less than 1 mg precious metal loading per cm^2. It is hoped that this concept will lead to cost reductions which will allow DMFC to be competitive for small-scale stationary applications in the medium term and for transportation in the long term.

Other concepts such as H_2-air solid polymer fuel cells are being investigated in F.R.G. (Siemens) and Italy (de Nora).

Work on alkaline fuel cells has been carried out in Europe for many years. Siemens developed a H2-O2 AFC of 7.5 kW with a mobile electrolyte. AFC are also used in military applications (submarines). Elenco (Belgium) has a small pilot production facility for H2-air AFC (2.5 MW/year) and is developing a hydrogen fueled AFC of 75 kW for integration in a bus for public transport in Amsterdam.

The European Space Agency has an extensive programme on AFC for space applications with 18 MECU for a period of two years (1988 - 1989) which is reported in more detail by Baron on p. 207.

Journal of Power Sources, 29 (1990) 143 - 148   143

FUEL CELLS IN ITALY

RAFFAELE VELLONE*

ENEA/FARE, 00060 S. Maria di Galeria, C.R.E. Casaccia, Via Anguillarese 301, Rome (Italy)

ANGELO DUFOUR

Ansaldo Ricerche, Genoa (Italy)

Introduction

Since the beginning of the seventies, scattered research activities on fuel cells were conducted in Italy producing a source of basic knowledge. In 1983 ENEA gathered all the potential operators (industries, utilities, users, research structures) into a working group aimed at verifying the feasibility of a national research and development (R&D) programme. The working group conducted a thorough evaluation of the potential benefits of FC technology in the Italian energy system; thus, in 1984 the 'Progetto VOLTA' was launched which established the guidelines for Italian activities in this field.

Starting from these guidelines, the Italian programme has been developed in the last five years: the up-to-date status is presented here.

Italian interests in fuel cells

The well known advantages of fuel cell plants (high efficiency, low polluting emissions, multi-fuel supply possibility) can fit the future needs of the Italian energy system: the strong dependence on imported fuels, mainly oil, is going to be reduced by saving energy and diversifying fuel sources. The growing energy demand in the densely populated urban areas could be better satisfied by dispersed-type power plants with acceptable environmental impact.

Market studies [1] allow quantification of the potential energy saving projected to 2010: according to a pure extrapolation of the present Italian energy scenario, 1 Mtoe/year could be saved. Introducing some hypotheses about the evolution of the energy scenario such as: (a) the government promotes industrial self-production of electricity by means of an adequate price policy, (b) FC plant cost is reduced to 700 $/kW, (c) combustion engine traffic is restricted in urban areas of historic interest, then the fuel

---

*Author to whom correspondence should be addressed.

0378-7753/90/$3.50

cell potential market increases and the figure of energy saving in 2010 goes up to 3.5 Mtoe.

The estimated market is illustrated in Table 1, for the 'extrapolative' scenario and for the 'evolutive' scenario. Accordingly, fuel cell technology is going to be an attractive industrial business so that another goal of the Italian programme is to encourage national industries to enter this field.

TABLE 1

Potential Italian market of fuel cell systems in 2010 [1]

Application	Installed power (MW)	
	Extrapolative scenario	Evolutive scenario
Power plants (1 - 100 MW)		
Electric utilities	1500	2500
Industrial self-producers		6200
On-site generators	200	500
(20 - 200 kW)		
Small generators	100	200
(1 - 10 kW)		
Electric vehicles	2500	6000
Total	4300	15400

Italian R&D activities

Phosphoric acid FC

The world leaders (U.S.A. and Japanese manufacturers) have reached a quite satisfactory stage of development in phosphoric acid stack technology, so that it has been decided to cease competing in this field, except for very small generators. However, PAFC demonstration plants are very important in the national programme strategy, because they are essential to promote both the internal market and the industrial experience in building up fuel cell plants.

The main effort in this strategy is represented by the 1 MW PAFC power plant to be installed in Milan. The plant will be designed and built up by Ansaldo, then AEM (Milan Municipal Energy Authority) will take charge in the operational phase, scheduled for 1991.

The electrochemical section (two 670 kW fuel cell stacks) is supplied by the International Fuel Cell Co. (U.S.A.) and the methane reforming section by Haldor Topsøe (DK). The conceptional design activity is nearly completed and Ansaldo is approaching the detailed design phase. The work undertaken so far has permitted the acquisition of very useful experience in the fuel cell system. Meanwhile, AEM is involved in the site preparation and licensing; all the relevant problems should be solved in a short time.

We are confident of a successful experiment for this first European FC power plant; this could be very favourable toward the commercialization of fuel cells.

Application of FC systems is on-site generation (20 - 200 kW) is very promising and should represent the first step in market penetration. Italy participates in the CEC 25 kW programme: one plant will be installed in 1990 and tested in Bologna at the city Municipal Energy Authority (ACOSER). The installation of a similar 25 kW plant, also constructed by KTI with Fuji cell stacks, is underway at ENEA Energy Research Center in Casaccia, near Rome. This prototype is a breadboard unit that will permit an extended characterization of the plant and its components.

An on-site demonstration programme is likely to be enhanced by testing one or more IFC 200 kW plants: some utilities are evaluating this opportunity.

The small portable generators programme, jointly established by ENEA and the Ministry of Defence, includes two different R&D aspects: (a) the cell stack development and (b) the system design, construction and test. The latter aspect is carried out by Ansaldo and Tecnars, and is aimed at manufacturing some compact generators of 1 and 5 kW equipped with cell stacks produced by a foreign supplier. Meanwhile, the Institute CNR-TAE of Messina is at an advanced stage in the development of a 1 kW stack entirely made with proprietary components and engineering. By now, the performance and endurance of monocells and 100 W stacks have been shown to be state-of-the-art. In a second phase of the programme, these two R&D aspects should be linked together to construct a portable generator equipped with the Italian stack.

Molten carbonate FC

A major goal of the Italian fuel cell programme is to develop molten carbonate FC stack technology, which is more promising than PAFC for several reasons (efficiency, cost, cogeneration capabilities, low sensitivity to fuel impurities).

In order to meet this goal the 'Progetto VOLTA' negotiated an international agreement with an experienced partner to acquire up-to-date knowhow and to carry out a joint development programme. Unfortunately, several problems have delayed this step, but we are very confident that a satisfactory collaboration will be arranged by the end of this year.

Notwithstanding, Italian activities in this field have grown considerably in the last five years, with contributions from industries and research centres.

Ansaldo have developed production techniques on a laboratory scale for both conventional and advanced components. Furthermore, Ansaldo bult and tested small cells up to 100 cm^2; performances were quite satisfactory and endurance tests lasted up to 2000 h. Studies of mechanisms of hot corrosion for metallic components were carried out by Ansaldo in collaboration with the Politechnic of Milan.

CNR-TAE of Messina accumulated an overall 30 000 h of endurance tests of components in 100 cm² IGT monocells with emphasis on modeling and investigation of the behaviour in the lowest temperature range (600 - 650 °C). This work, coupled with in depth post-mortem analysis of all components, has clearly indicated that extended endurance under lower T conditions will outweigh inherent losses in power. A catalyst suitable for internal reforming has been developed and prepared for extensive in-cell experimentation.

CISE's activities have been focused on the development of novel cathodes through studies of the mechanisms of their dissolution under simulated conditions.

Part of the activities of Ansaldo, CNR-TAE and CISE are carried out within the framework of a CEC research programme in this field. The University of Genoa developed mathermatical models for components, cell and stack. ENEA used a three electrode cell in order to carry out the electrochemical characterization of components and small cells. The use of molten carbonate fuel cells in coal power plants was studied by CRITA, from both the technical and economic point of view, with reference to the Italian energy system.

All these activities have generated experiences which are very useful for the larger programme that is going to start shortly.

Solid polymer electrolyte FC

In consideration of the recent progress and the growing interest for this technology, an R&D programme was established in 1988 within the framework of cooperation between ENEA and De Nora. The goal of this programme is the construction and characterization of a 10 kW stack in 1990.

Solid oxide FC

Basic research activities on materials and fabrication techniques of components were carried out in the past by CISE and CNR-IRTEC (Ceramic Research Institute), in order to investigate the critical problems of this technology.

A larger project is starting now, in close connection with the CEC programmes in this field and with the objective of building and evaluating cells and small stacks. A variety of materials and fabrication procedures will be investigated in order to develop systems made by processes easily amenable to mass production and that at the same time guarantee as high efficiency and power density as possible.

The main organizations involved are CISE, ENEA, Eniricerche, Milano Ricerche and National Research Council (CNR, with its Institutes TAE, IRTEC and CSTE).

FC system for transportation

The application of fuel cell technology for transportation is very promising; an integrated programme in this field is under evaluation. It will include the complete development of both the fuel cell generator and the electric vehicle, involving users and diverse industrial expertise.

Costs and sources of financing

Table 2 reports the distribution of funds among the various activities of the Italian programme.

The main share of these funds is devoted to design, bulding and testing of the Milan 1 MW PAFC plant; the necessary 40 billion Lire (about 28.5

TABLE 2

Total funds in the Italian fuel cell programme (values in million Italian Lire)

	Appropriation of funds until 1988	Expenditures until 1988	Funds to be engaged in 1989 - 1992
Phosphoric acid FC			
1 MW demonstration plant	40625	2764	
Small transportable generators	7500	1610	1000
On-site generators	3980	600	2500
Molten carbonate FC	7482	4860	27000
Solid oxide FC	715	565	5000
Polymeric electrolyte FC	1972	220	2000
FC systems for transportation			4000
Market and application studies	682	540	1000
Total	62956	11159	42500

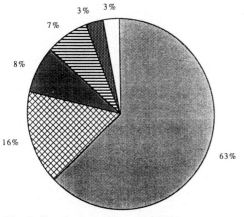

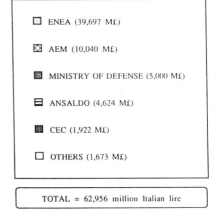

ENEA (39,697 M£)

AEM (10,040 M£)

MINISTRY OF DEFENSE (5,000 M£)

ANSALDO (4,624 M£)

CEC (1,922 M£)

OTHERS (1,673 M£)

TOTAL = 62,956 million Italian lire

Fig. 1. Funds engaged until 1988.

million US$) are financed by ENEA ($\approx 63\%$), AEM ($\approx 25\%$), Ansaldo ($\approx 10\%$), CEC ($\approx 2\%$).

Another point to be emphasised is that future appropriation should mainly concern molten carbonate FC technology.

The distribution of all funds committed until 1988 among financial sources is shown in Fig. 1.

Reference

1 TESI, Institute Ricerche Breda, AEM, under ENEA contract.

Journal of Power Sources, 29 (1990) 149 - 166

THE DESIGN OF ALKALINE FUEL CELLS

K. STRASSER

*Siemens AG, UB KWU, U9 261, Hammerbacherstrasse 12 + 14, P.O. Box 3220,
D8520-Erlangen (F.R.G.)*

Introduction

Since the beginning of the sixties different fuel cell systems have been
investigated. They are characterized by using different
- Fuels
- Oxidants
- Electrolytes
- Catalysts
- Electrodes
- Cell design

Fuel cells with technical pure hydrogen or reformer gas as the fuel
and oxygen or air as the oxidant are the most important. One of the different types — namely the low temperature fuel cell — may be applied as the
electrical power source of a power storage system. This system is not suitable
for power plant application, to which middle and high temperature fuel cells,
i.e. PAFC, MCFC and SOFC are more adapted.

Low temperature fuel cells are cells with alkaline electrolyte and solid
polymer electrolyte and have been approved as the electrical power supply
in the U.S. Space programs Gemini and Apollo.

The following properties are advantageous:
- Overall efficiency is not limited by the Carnot factor
- Overall efficiency increases at part load
- Energy density is distinctly higher than that of the accumulator
- Series and parallel connections of the modules make good redun-
dancy
- The fuel cell operates noiselessly, in the case of operating with
technical pure reactants, without polluting gases, at low maintenance.

The typical principles of low temperature fuel cells

The operating principle of the low temperature fuel cell with mobile,
alkaline electrolyte is shown Fig. 1. In the reaction hydrogen is oxidized at
the anode, while oxygen is reduced at the cathode. The reaction products
are water and heat, which dilute and heat up the liquid electrolyte and are
removed from the cell by the electrolyte circulation. Usually 30% potassium
hydroxide is used as the electrolyte.

0378-7753/90/$3.50

150

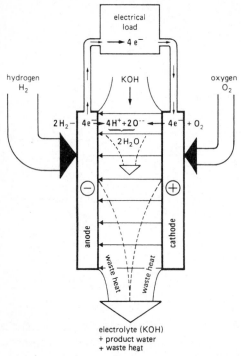

Fig. 1. Principle of the alkaline fuel cell with mobile electrolyte.

The principle of the second type of low temperature fuel cell, the alkaline fuel cell with base electrolyte is depicted in Fig. 2. In this case the electrolyte is based in a porous matrix for example in asbestos. The product water is removed from the cell by a hydrogen loop as water vapor, the waste heat by a coolant circulation. Using thin electrodes an additional porous sheet in the cell is necessary to store the product water during the heat up phase.

The principle of the third type of low temperature fuel cell, the PEM, fuel cell with polymer electrolyte, which is similar to the alkaline matrix cell, is shown in Fig. 3. The electrolyte of this cell is from a proton exchange material. Liquid water is produced at the cathode side and is removed from the cell by an oxygen loop or by static water management. Waste heat is directly removed by a coolant circulation.

The theoretical value of the cell voltage — dependent on the upper heat value of the hydrogen — amounts to 1.48 V. Measurable cell voltage at open circuit is a little higher than 1 V. As Fig. 4 shows, usable cell voltage decreases by increasing current, while voltage losses at low currents are caused primarily by the polarization of the cathode. At higher currents ohmic resistance increases and finally at high current, transport losses become significant. Usually transport losses at rated current may be neglected. Rated current depends on different operating conditions as well as on the fuel

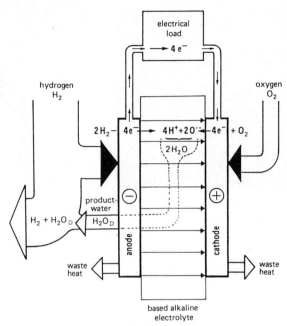

Fig. 2. Principle of alkaline matrix cell.

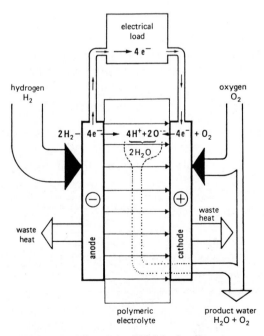

Fig. 3. Principle of the PEM fuel cell.

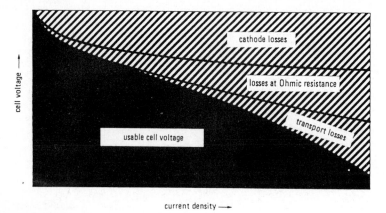

cell voltage

cathode losses

losses at Ohmic resistance

transport losses

usable cell voltage

current density ⟶

Fig. 4. Inner losses of cell voltage (schematic).

cell type, the electrode type or cell design. A current density of 0.5 A/cm^2 has been reached in complete fuel cell systems and more than 1.0 A/cm^2 in experimental cells.

Basic design of the fuel cell

Low temperature fuel cells are under development at different companies. Table 1 shows an overview of this activity.

An example of the basic design of the alkaline fuel cell with mobile electrolyte, as developed at Siemens, is shown in Fig. 5.

By the cell design the liquid electrolyte is separated from the gaseous reactants by asbestos diaphragms. The electrodes — anode from Raney-Nickel, cathode from doped silver — are connected uniformly to the electrolyte as well as to the current collecting side by pressing isostatically with pneumatic pressure cushions.

The active area of the cell is 340 cm^2. The single parts are sealed to each other by cementing as well as by an elastomer frame. The total network of the supplying channels is formed from the holes in the frames by mounting the stack.

At an operating temperature of 80 °C and a reactant pressure of about 2 bar a, the fuel cell can be loaded with a current density of 400 mA/cm^2 at ~0.8 V.

Figure 6 shows an expanded view of the cell.

The basic design of the Elenco fuel cell is not bipolar. It is edge-connected and stacks are assembled from modules.

Figure 7 shows an example of the matrix-type used at Siemens. The cell design is very similar to the cell with mobile electrolyte, as described. In principle there is no difference in the design of the matrix cell and the PEM cell. In Fig. 8 current voltage characteristics of Siemens fuel cells with a

TABLE 1

Current work on low temperature fuel cells

Type of fuel cell	Company	Application	Characteristics
Fuel cell with alkaline, mobile electrolyte	Siemens, F.R.G.	underwater application	H_2/O_2; about 2 bar a*; 80 °C Ni-anode/Ag-cathode
	Elenco, Belgium	electric vehicle (Bus)	H_2/air; low pressure anode and cathode with low Pt-loading
Fuel cell with alkaline matrix electrolyte	IFC, U.S.A.–Japan	space application	H_2/O_2; about 4 bar; 80 °C anode and cathode with high Pt-loading
	Siemens, F.R.G.	space	H_2/O_2; different conditions different electrode material
PEM fuel cell (for comparison)	Siemens, F.R.G.		H_2/O_2; about 2 bar a; 80 °C
	BTC, Canada		H_2/O_2; H_2/air; Methanol/air
	LANL, U.S.A.	different applications	H_2/O_2; H_2/air; Methanol/air
	GM, U.S.A.		Methanol/air
	Energenics, U.S.A.		H_2/O_2
	IFC, U.S.A.		Methanol/O_2

*a, absolute.

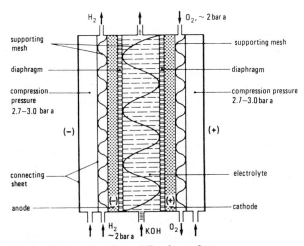

Fig. 5. Fuel cell with mobile electrolyte.

154

Fig. 6. Expanded view of the Siemens FC.

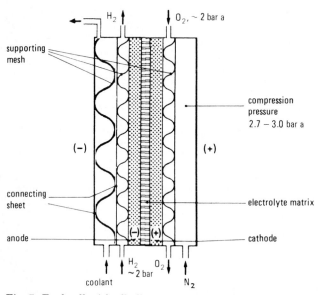

Fig. 7. Fuel cell with alkaline matrix electrolyte.

nickel anode and a silver cathode at different operating conditions are depicted.

The fuel cell system as developed at Siemens

To operate fuel cells, independent of the type, different components are necessary.

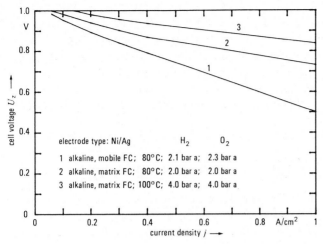

Fig. 8. Current voltage characteristics of Siemens fuel cells.

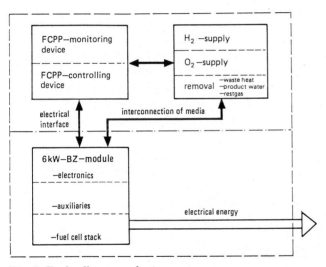

Fig. 9. Fuel cell power plant.

The main components of the system are shown in Fig. 9.
- Fuel cell modules
- H_2 supply
- O_2 supply
- Device for removing waste heat, product water and restgas
- Installation for controlling and monitoring

The main components of the FC system are connected at the electrical interface and the interconnection of the different media. Because of the modular design, the adaption of the fuel cell system to special requirements of the application is relatively simple.

The FC module

The FC module is assembled from a 60-celled FC stack, the electrolyte evaporator stack, the electromechanical and electronic control unit.

For stack operation auxiliaries are necessary, which have to be assigned to the FC module (Fig. 10). They envelop the following device for

- Supplying with H_2, O_2 and N_2 gas
- Removing product water and waste heat from the fuel cells via electrolyte circulation
- Separating product water and waste heat from the electrolyte by a slit-type evaporator
- Removing waste heat from the electrolyte evaporator by a coolant circulation

The FC module block is incorporated in a pressure tank. It is operated at 3 bar a with N_2 as a protecting atmosphere, preventing a mixture of reactants in the surroundings by leakages, and has to be supplied with defined static pressures for H_2, O_2, N_2 and coolant.

Figure 11 shows the technical data for the FC module at defined conditions, and the outline dimensions.

Removal of product water and heat

The principle applied in the electrolyte regenerator for water and heat removal is shown in Fig. 12. The spaces through which electrolyte flows alternate with those containing coolant. A gas-filled gap separates each electrolyte and coolant space. The electrolyte spaces are surrounded with asbestos diaphragms and the coolant spaces with thin, non-porous sheets.

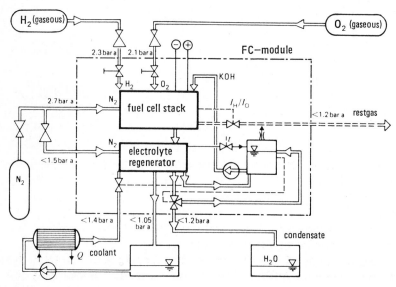

Fig. 10. Fuel cell power plant, consisting of the FC module, pressure gas bottles and coolant circulation.

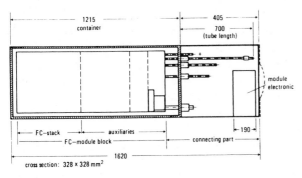

power   6 kW
voltage   46–48 V
efficiency   61–63%
efficiency at ~20% load   71–72%
temperature   ~80° C
H_2–pressure   2.3 bar a
O_2–pressure   2.1 bar a
dimensions   0.328 x 0.328 x 1.62 m^3
weight, total   215 kg
weight of the container   125 kg

Fig. 11. Technical data of the 6 kW FC module BZA 4-2.

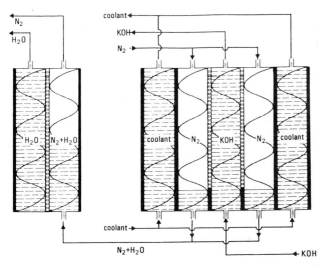

Fig. 12. Principle of the water and heat removal system in FC module BZA 4-2.

The product water is removed by evaporation and condensation, caused by the different water vapor partial pressures at the electrolyte-side and coolant-side surfaces. The static pressure in the gas gap and a pressure lock aid removing the condensate.

Removal of inert gas

The principle applied in removing inert gases from the fuel cell is depicted in Fig. 13. H_2 and O_2 are streamed through the FC stack in opposite directions in a cascading manner. Inert gases are collected in the last step of the cascading system. This is the gas space in a cell which is electrically connected in parallel with another cell, as shown in Fig. 14.

This configuration is responsible for the fact that the load current in a cell changes when the inert gas concentration in the gas space in the cell is increased. Current sensors record this deviation. An electronic device switches on a valve at the stack outlet, releasing the inert gases, when the current through the H_2 reference cell is decreased to 95%, and through the O_2 reference cell to 80%.

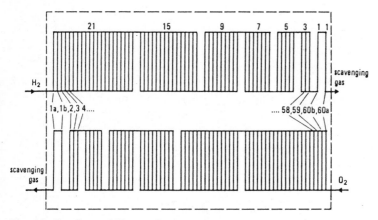

Fig. 13. Gas flow of H_2 and O_2 in FC module BZA 4-2.

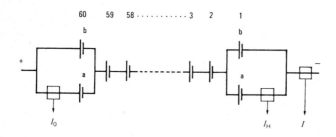

purging conditions: O_2: $2I_O/I = 0.80$

H_2: $2I_H/I = 0.95$

Fig. 14. Electrical connection of the fuel cells and location of the current sensors for current controlled purge gas removal in FC module BZA 4-2.

Fig. 15. FC module: 48 V, 125 A, 6 kW.

Controlling and monitoring device

Beside the four controlled functions
- Electrolyte temperature
- Electrolyte concentration
- H_2 inert gas removal
- O_2 inert gas removal

the following functions are monitored
- Module voltage
- H_2 pressure in the FC module
- Voltage of the H_2 reference cell
- Voltage of the O_2 reference cell
- Electrolyte temperature
- Electrolyte volume
- Electrolyte flow

The module is automatically shutdown when a monitoring circuit responds. Figure 15 shows a complete connected module.

The complete module with the pressure housing removed and the tubing also connected to the supply interface is shown in Fig. 16. For the FC stack in the background, which is the energy converter, only 15% of the total module volume is needed.

Operational behaviour and performance data

Static load behaviour

The operating behaviour at static or dynamic loading has been investigated for many FC modules. The measurements are related to more than

160

Fig. 16, FC module without pressure container, supplying tubes connected at the interface.

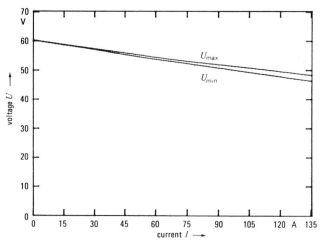

Fig. 17. Current/voltage characteristic at rated operating conditions after 100 h of operation in FC module BZA 4-2.

1000 cells of the above-mentioned type. A typical current/voltage characteristic of the FC module is shown in Fig. 17. At rated current 46 - 48 V are attainable.

The spread in mean cell voltage as a result of variation in electrolyte temperature and as-manufactured deviations lies between the U_{max} and U_{min} curves and is equal to about ±15 mV at a rated current of 135 A. Figure 18 shows the voltage of the individual fuel cells at rated current. The voltage in the outermost cells is higher than in the other cells because connection in parallel involves lower loads (see Fig. 14).

The power per module under rated operating conditions is plotted against load current in Fig. 19 and amounts to 6 kW rated power. Short-term overloads are possible. The magnitude of the load current is limited by a lower voltage limit and the duration of the overload by the amount of heat removed. Both limits are monitored within the module.

Load currents below 20 A are allowable for only limited time periods because they interfere with the removal of water from the cells and with electrical potential. No problems are encountered for time periods of 5 min.

The module can be operated with 'technical pure' gases, 99.5% oxygen and 99.95% hydrogen. The impurities are separated by the installed inert

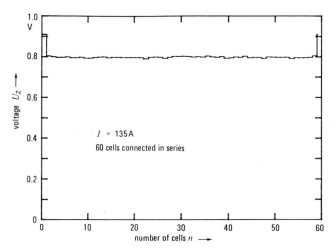

Fig. 18. Cell voltages at rated operating conditions in FC module BZA 4-2.

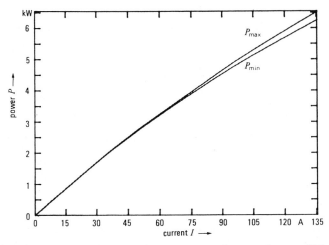

Fig. 19. Current/power diagram at rated operating conditions after 100 h of operation of FC module BZA 4-2.

162

removal system. About 80% of the rest gases are impurities of the reactants as shown in Fig. 20. This figure shows that this method is considerably more efficient than the others, especially when the attempt is made to get the largest possible amount of reactants to participate in the reaction and to have the smallest possible amount of residual gas left over. The curve of the overall efficiency, which corresponds to the static loading characteristic, is shown in Fig. 21. The efficiency is related to the lower heating value of the hydrogen. Only the power consumption of the coolant circulation has not been taken into account. The efficiency amounts to 61 - 63% at rated current and the maximum at part load to 71 - 72%. These numbers are nearly a factor of two higher than the efficiency of thermal engines.

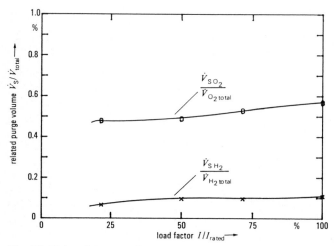

Fig. 20. Related purge volume as a function of the load factor of FC module BZA 4-2.

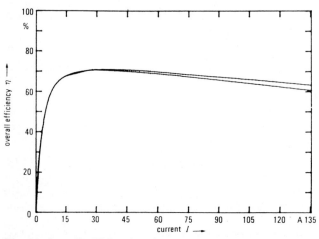

Fig. 21. Overall efficiencies after 100 h of operation of FC module BZA 4-2.

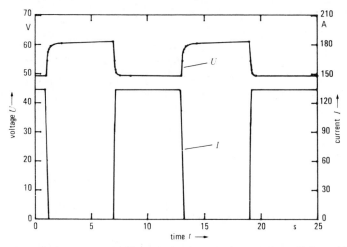

Fig. 22. Current and voltage behaviour under rated conditions of FC module BZA 4-2.

Dynamic load behaviour

As an energy converter out of operation the fuel cell module is without voltage. Regardless of the temperature of the module, it reaches the voltage dictated by the load current and electrolyte temperature within 5 s of being turned on. The increase in voltage is dependent only on the speed of gas exchange in the fuel cells. In principle, the fuel cell can be started up under load. During start-up time about 4 l of gas mixture (H_2/O_2) are blown into the residual gas system.

Using its own waste heat, it takes about 15 min at a constant module voltage of 48 V to heat up the module from room temperature to operating temperature of 80 °C. The lower the load, the longer the heating up phase. When the load current is altered, the module voltage follows the load spontaneously, *i.e.* in less than 100 ms (Fig. 22).

In the event of a short in the load circuit, the internal voltage monitoring function shuts down the module. Figure 23 shows a plot of current and voltage in the cell over time for this case. It can be seen from the plot that the maximum short circuit current is 1.300 A and the fade-out is about 1 s. The module can also be operated for up to 5 min when declined ±45° to the transverse or longitudinal axis. There is no time limit on operation when the module is tilted less than ±20°.

While switching off the module is disconnected from the load circuit, the O_2 supply interrupted and the energy stored in the fuel cells discharged across a 2 ohm resistor. Within about 25 s, the module is discharged to the point where the voltage remaining is less than 10% of the nominal. Upon shutdown about 10 l of gas mixture (H_2/O_2) are blown into the residual gas system.

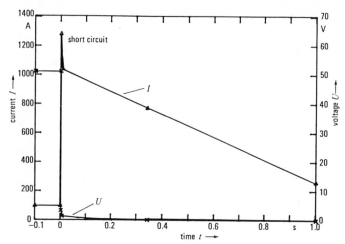

Fig. 23. Short circuit behaviour starting from 100 A at operating conditions of FC module BZA 4-2.

Demonstration of a FC system as a power source

To demonstrate and to investigate the function and the reliability of the FC system a power source was constructed, which consists of eight FC modules of the described 6 kW-type, mounted into a rack, the H_2 and O_2 supply and removal of the waste products and the central electrical installation. Total output of the power source is about 50 kW. Each set of four modules is connected in series, providing the required 192 V total voltage at about 250 A. The setup of the fuel cell power plant is shown in Fig. 24.

The modules are connected at the supply interface and are supplied in parallel. The overall electrical system controls and monitors the operation. Figure 25 shows the fuel cell power plant in test operation. The required data and the high system reliability were demonstrated during more than 20 000 h of accumulated module operation.

The time history of module voltage (Fig. 26) for 4 modules demonstrates how slowly the system deteriorates with operating time. Similar results were obtained in tests on cells over several thousand hours of operation.

The proper functioning and high reliability of the total power source of 100 kW was demonstrated during last year, when tests as a power source of an air independent propulsion system on a German submarine were carried out with very good results. The tests were finished in February of 1989.

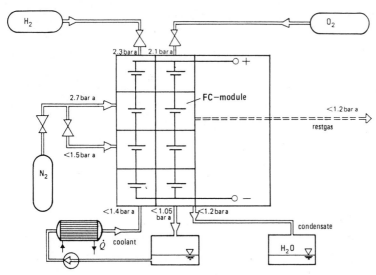

Fig. 24. Function schematic of FC system.

Fig. 25. FC unit consisting of 8 modules: 192 V, 250 A, 48 kW.

Conclusions

From the technical point of view alkaline fuel cells have proved satisfactory even with mobile and matrix electrolytes. The modular designed fuel cell system, as described by Siemens, made it possible to demonstrate its advantages in actual operation as the 100 kW power source for an air independent propulsion system. This may be the largest power source with alkaline fuel cells to date.

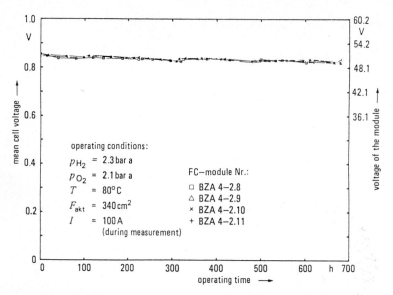

Fig. 26. Long term behaviour of FC module BZA 4-2.

On this basis, Siemens is now starting the development of a power source for the European space program. It will be an alkaline fuel cell of the matrix type.

The characteristics of cost, technical maturity and system advantages will permit broader application. As special requirements can be satisfied, the system may at first be applied in these special applications.

Bibliography

K. Strasser, L. Blum and W. Stühler, An advanced alkaline H_2/O_2 fuel cell assembly as a compact module of a 50 kW power source, *Fuel Cell Seminar '88, Long Beach, CA, U.S.A.*

Short Contributions

Journal of Power Sources, 29 (1990) 169 - 179

MARINE AND NAVAL APPLICATIONS OF FUEL CELLS FOR PROPULSION: THE PROCESS SELECTION

WILLIAM H. KUMM

Arctic Energies Ltd., 511 Heavitree Lane, Severna Park, MD 21146 (U.S.A.)

The system analysis steps and the tradeoff considerations governing the process selection for the marine application of fuel cells to propel large ships is described. The analytical process was used, in part, during 1987 in the performance of a study of a fuel cell propelled combatant ship conceptual design under a U.S. Navy contract [1]. Figure 1 shows the marine fuel cell process selection procedure. Eighteen steps are involved in the surface ship case. Steps 19 and 20 would also be involved in evaluating fuel cells to power submarines. The steps are discussed in sequence.

1. Type

Six types of fuel cells are considered in the analysis. They are:
(A) Alkaline
(B) Proton exchange membrane or solid polymer electrolyte
(C) Phosphoric acid: (a) external reformer, (b) internal reformation
(D) Super acids
(E) Molten carbonate
(F) Solid oxide
Alkaline fuel cells have been used in the NASA space program in the Apollo system and the Space Shuttle system. The others have received varying degrees of DOD, DOE, EPRI and GRI research and development funding. In 1985/1986 the Office of Technology Assessment (OTA) of the U.S. Congress completed a Technical Memorandum on the Marine Applications for Fuel Cell Development [2]. The author participated in the preparation of the OTA report. The report provides an excellent general treatment of the subject. The foreword of the OTA report says: "To date almost no attention has been given to the potential marine applications for fuel cell technologies. Nevertheless, some of the benefits that fuel cells may offer to the utility industry may also apply to some marine uses.".

The types of fuel cells considered for marine applications which are listed above (A through F) are in ascending order of operating temperature. As the surface ship sequence of 18 process selection steps is carried forward, various types are dropped from further consideration because their charac-

0378-7753/90/$3.50

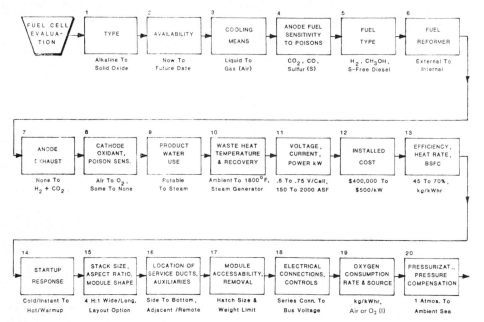

Fig. 1. Marine fuel cell process selection.

teristics are inappropriate to the marine applications, which are typically at multimegawatt power levels.

2. Availability

The alkaline, proton exchange membrane and phosphoric acid types are available now. The molten carbonate type is classed as being available in the near future. Both the super acids and the solid oxide types are in the further future and appear to offer little advantage over the phosphoric acid and the molten carbonate types respectively, for marine application.

3. Cooling means

The primary coolant can be liquid or gaseous. However, the liquid must not conduct electricity. This typically means demineralized fresh water or a dielectric liquid such as mineral oil. However, above the temperature of the proton exchange membrane fuel cells at 180 to 200 °F the coolant water will actually be in two phase flow in phosphoric acid or higher temperature fuel cells. Thus the use of gas cooling, typically air, is common and provides lower system weight, no liquid leakage and ease of access to the fuel cell stacks. The two highest temperature type fuel cells, molten carbonate and solid oxide, are both gas (air) cooled.

One of the main reasons the alkaline fuel cells are not recommended for ship propulsion power use is their total intolerance to carbon and air contaminants. This fuel cell technology was designed for spacecraft. In that weight-critical and largely cost-unlimited application alkaline fuel cell power plants work well.

4. Anode fuel sensitivity

Many readily available carbonaceous liquid fuels contain contaminants. The fuel cell types listed in Section 1 vary from those with no tolerance (alkaline) to those with some tolerance for the most common and pernicious fuel poison, sulfur. Other fuel contaminants include carbon dioxide and carbon monoxide, both of which are formed in the thermochemical reformation of carbonaceous fuels to hydrogen, which is the fuel actually used by the fuel cells. Alkaline and proton exchange membrane fuel cells are permanently poisoned by carbon monoxide and alkaline fuel cells are poisoned by carbon dioxide. Thus these two fuel cell types are not readily fed from reformed hydrocarbon fuels. Proton exchange membrane fuel cells can be used with reformed sulfur-free fuels such as methanol so long as great care is taken to shift any carbon monoxide produced over to carbon dioxide. The phosphoric acid and super acids fuel cells have complete carbon dioxide tolerance and a degree of carbon monoxide tolerance.

Molten carbonate fuel cells can operate without carbon monoxide tolerance problems but are still affected by sulfur. Only the extremely high temperature solid oxide type has some sulfur tolerance. This effectively means that whatever fuel cell type is contemplated for marine or naval uses the fuel must be sulfur-free. This is most readily done in the U.S. by having the Federal Fuel Supply System specify and stock sulfur-free diesel fuel for marine use on fuel cell powered ships.

5. Fuel type

The sulfur-free fuels for the fuel cell types listed in Section 1 are, in order: pure hydrogen only for the alkaline fuel cells; through hydrogen or methanol for proton exchange membrane; to hydrogen, methanol, natural gas and sulfur-free diesel for phosphoric acid, super acids and molten carbonate fuel cells. The degree of sulfur tolerance of the high temperature solid oxide type makes it (prospectively) the only type which might be able to use ordinary diesel fuel, which contains some sulfur. The heavy bunker C type of marine fuels also have traces of heavy metals such as vanadium, and the desulfurization step could remove such impurities as well.

As a marine fuel sulfur-free diesel has all the characteristics of marine diesel fuel such as high heating value, *i.e.* Btu/lb, high flash point and common usage familiarity in a variety of existing marine power plants of the heat engine type. As major users of such fuel the U.S. federal agencies such as the Coast Guard, NOAA, Army Corps of Engineers and the Navy could secure the sulfur-free diesel at little cost increase over existing diesel fuel. The sulfur in existing marine diesel fuel does the machinery no benefit and in fact reduces equipment life, particularly exhaust stacks and heat recovery systems. This is due to sulfur and hydrogen sulfide, which in the presence of steam, forms corrosion products such as sulfuric acid.

Methanol is an excellent clean liquid fuel. It reforms readily to hydrogen plus carbon dioxide at a relatively low temperature. It has a lower flash point than diesel fuel. However, a methanol fire onboard a ship can be extinguished with water because methanol and water are miscible. The major system level disadvantage of methanol for the surface ship case is the low heating value, *i.e.* Btu/lb, compared to diesel fuel or even jet fuels such as JP5. As is discussed further below in Section 13, this is an endurance tradeoff matter. If ship's tanks can carry adequate loads of methanol for the mission endurance then methanol fuel is an excellent choice. However, a sulfur-free diesel fueled ship will travel substantially further on the same tank fill.

6. Fuel reformer

With the exception of the alkaline fuel cell type, which must use pure hydrogen fuel, all the others can use a reformed fuel. The temperature at which methanol can be thermochemically reformed is in the order of 770 °F. If the fuel cell stack temperature is less than that there has to be an external heat source, or some of the fuel must be burned, to provide the needed temperature for the catalytic function to occur. Proton exchange membrane fuel cells must burn some methanol at the reformer, or permit some of the anode tail gas hydrogen to return to the reformer to be used as its fuel. The same is true of the phosphoric acid or the super acids type fuel cells except that the thermal integration is better and less fuel is burned in the reformation function.

In the case of the phosphoric acid fuel cells there is a further design variant which is possible, which provides a better thermal integration between the thermochemical and electrochemical functions. It involves raising the fuel cell operating temperature to about 450 from 375 °F, with a corresponding reduction in cell voltage of about 10%, and a probable reduction in the service life of the same order of magnitude. In this configuration the thermochemical function would be performed in modified 'cooling passage' plates, located between each group of 5 cells (called a 'substack'). These 'reforming plates' would have appropriate catalysts in their passages and by feeding methanol vapor plus water vapor into their inputs the reformation would be carried out within the fuel cell stack. This type of internal reformation phosphoric acid fuel cell system will only work on methanol fuel because the reformation temperature is low enough. It would not be possible using diesel or other heavy fuels, which require much higher temperatures for reformation to hydrogen plus carbon dioxide. Internal reformation can reduce cost, weight and volume by as much as 25%, plus increasing energy conversion efficiency, so the approach is promising. It is discussed further below.

The big system-level tradeoff advantage of the molten carbonate fuel cell approach is that it can internally reform desulfurized diesel fuel with high efficiency, because it has a 1200 °F operating temperature. The high quality waste heat can also be used for other purposes onboard, as is discussed in Section 10. Solid oxide fuel cells appear to have the same generic advantage as the molten carbonate but they are further in the future.

7. Anode exhaust

In the case of the alkaline fuel cell there is no anode exhaust because all the pure hydrogen is consumed in making water and electricity. In the proton exchange membrane, phosphoric acid and super acids fuel cells there will be some unused hydrogen and carbon dioxide in the anode exhaust. In solid oxide fuel cells the product water (in the form of steam) is formed at the anodes and therefore exits via the anode exhaust manifold along with some unused hydrogen and carbon dioxide.

8. Cathode oxidant

The only oxidant acceptable to alkaline fuel cells is pure oxygen. The proton exchange membrane, phosphoric acid and super acids fuel cells can all operate from air or oxygen. At sea the salt must be removed from the sea air but standard marine air purification and handling technology is available to remove salt from the air.

The oxidant requirements of the two high temperature fuel cell types, molten carbonate and solid oxide, are readily met with air.

9. Product water use

All fuel cells produce fresh water. In ascending order of operating temperature from the alkaline type, the water produced in the fuel cell stacks will be predominantly in the form of steam. The heat carried in the superheated steam from the two high temperature fuel cell types (molten carbonate and solid oxide) can be used directly or indirectly through a heat exchanger, to operate a 'bottoming plant' such as a heat engine to generate additional electricity. Once the steam has been condensed it is then available for use as potable water for hotel load needs onboard.

10. Waste heat

As has been previously explained in Section 1, the various fuel cell types were shown in ascending order of operating temperature, from the low temperature alkaline type to the very high temperature solid oxide type. Only three of the fuel cell types (C(a), C(b) and E) are considered from this point on as appropriate for near term consideration as candidate ship power plants. Only the molten carbonate 1200 °F operating temperature type can be considered for thermal integration at the ship services level. The 375 to 450 °F temperature of the phosphoric acid type fuel cells, whether using external or internal fuel reformation of methanol, only permits their waste heat to be used for fuel cell related functions such as methanol preheating, i.e. vaporization and condensed product water revaporization to feed the steam reforming function. The 1800 °F solid oxide (type F) may well prove to be applicable in a later time period, as may other future fuel cell types.

Waste heat removal can be by the use of air, other gasses in primary coolant loops, demineralized fresh water or a dielectric liquid such as mineral oil. Air cooling provides the lowest weight per installed kW of fuel cell power plant. All the high temperature fuel cell stacks are air or gas cooled.

11. Amperes per square foot and volts

These two fuel cell electrical characteristics define the power output possible from typical fuel cell stacks. For the proton exchange membrane recent work has demonstrated substantial increases in the current capacity in A/ft^2. Values of 2000 A/ft^2 represent significantly higher (by a factor of 10) current capacity than the phosphoric acid or molten carbonate fuel cells. The cell voltage is lower for the proton exchange membrane type, at 0.5 V as compared to 0.6 for phosphoric acid or 0.75 for molten carbonate, but the current capacity factor still dominates in the calculation of the power output of a typical stack, per unit of volume or weight. These considerations make the less energy conversion efficient proton exchange membrane type attractive for weight-limited system applications, but not necessarily for the large installed power level case of surface ships, as is discussed further below, in Section 15.

12. Installed cost

The matter of capital cost is customarily described in terms of $/kW installed, which refers to a power plant with all of its auxiliaries, unless otherwise stated. Certain shipboard services such as air handling blower systems and heat exchangers can be shared among a number of fuel cell power plant modules. Power conditioning, used typically to convert the fuel cell stacks's d.c. output power to a.c., can also be shared by multiple fuel cell power plant modules. An individual module must be able to start up from cold, from a hot stand-by condition, or with preheaters over a brief start-up time.

Within a fuel cell power plant module, such as a present day methanol fueled phosphoric acid type, the stack represents about 67% of the cost, the external reformer 23% and the control system 10%. In the case of the internally reforming molten carbonate fuel cell type the combined stack represents 90% of the cost and the control system 10%. Typical capital costs in mass production for both phosphoric acid and molten carbonate are expected to be in the order of $600* in 1988 dollars. In small quantities, *i.e.* using model shop type assembly, the price per kW could be 5 to 10 times that figure. $500 per installed kW figure is comparable to the installed cost for a diesel electric generator, so that cost figures are close to equivalence, once fuel cells are in mass production.

13. Efficiency, heat rate, BSFC

The energy conversion efficiency of the marine fuel cell system is paramount. Because the endurance or range of the ship is controlled by the size of the fuel tankage, it follows that the lower the rate at which fuel is used, the greater the range. Typical heat engine ship propulsion power plants are 35% efficient. Fuel cell power plants can have efficiencies of a much as 70% on air and even higher in submarine applications on pure oxygen, as discussed in Section 19. The derivation of the 70% efficiency figure for a marine molten carbonate installation is as follows: 55% for the direct energy conversion; 5% is for the potable water credit because a separate fuel-consuming water production system would not be required. A waste heat recovery system would be capable of extracting a further 10% of the energy from the high quality exhaust heat from these fuel cells. The total power plant energy conversion efficiency is thus 70%.

The electric utility industry normally refers to the heat rate of energy converters such as steam generators. The marine equivalent measure is the fuel rate or the brake specific fuel consumption (BSFC), which is expressed in lbs of fuel per shaft horsepower hour, or more conveniently for electric drive ships (and submarines), as kilograms per kilowatt hour (kg/kW h) or metric tons per megawatt hour (tonnes/MW h).

The following calculation shows the relationship between the efficiency of the fuel cell, the heating value of the fuel and the resulting fuel rate or BSFC. The fuel in the calculation is diesel fuel with a heating value of 18 500 Btu/lb.

$$\text{Fuel rate in lb/kW h} = \frac{3413 \text{ Btu/kW h}}{70\% \times 18\,500 \text{ Btu/lb}} = 0.26$$

0.26 lb/kW h/2.2 lb/kg = 0.118 or 0.12 kg/kW h.

---

*Excluding inverter which is not envisaged for marine applications.

Had the fuel been methanol, with a heat content of 9550 Btu/lb, the fuel rate would be about doubled, to 0.23 kg/kW h. Thus, there is a real ship system-level endurance advantage to using diesel fuel rather than methanol, in a high temperature fuel cell power plant which permits the internal reformation of the fuel, while exhibiting the highest power system energy conversion efficiency.

14. Start-up response

This factor is important in some applications which may require instant start-up from cold conditions. The proton exchange membrane type has the best start-up time, if started on pure hydrogen, as it can operate cold. If a methanol fuel and reformer are used it is the reformer start-up which will govern the warm-up interval.

For the phosphoric acid and for the molten carbonate type, there is a preheat cycle of 10 to 15 min because the stacks must be at the operating temperature for net power to be produced. If a methanol external reformer is involved for the phosphoric acid fuel cell power plant there is also a comparable thermal start-up delay. However, the rapid cold start feature is not a characteristic that a large surface ship propulsion power plant needs to have.

From a hot stand-by condition to 'full throttle' there is some reformer lag on the part of large high temperature fuel cell power plants. However, this lag is of the same order as for a heat engine, and is thus not significant for any large surface propulsion power control condition.

15. Stack, size, aspect ratio, module shape

The proton exchange membrane technology permits large values for the current capacity in A/ft^2. The ship's electric bus voltage, the onboard physical layout considerations (such as deck height), or both, set the fuel cell stack heights. The voltage of each fuel cell module, such as 150 to 200 V, will thereby be defined. From this the A/ft^2 figure will define the cross-sectional area of the stack. The multiplication of stack height and stack cross-sectional area defines the form factor of the cube. In general, for the same power output the phosphoric acid and molten carbonate fuel cell power plant stacks will have the same shape and form factor. Proton exchange membrane fuel cell power plant stacks of the same power level would tend to be 'tall and skinny' because of the lesser cross-sectional area required due to the higher current capacity. As it is wise not to exceed an aspect ratio of height to either dimension of the base of a stack by more than 3 or 4 to 1, the form factor issue is of some importance when dimensioning power plant stacks and hence modules. This factor is also important in evaluating the shock loading response of fuel cell stacks. The closer the aspect ratio is to unity the more robust the stack will be.

16. Location of service ducts, auxiliaries

Modern ship design is increasingly based on replaceable power plant modules and reducing the shipyard time and port turnaround times by rapid removal and change out of such modules. Fuel cell power for ships is modular in nature so the choice becomes one of how best to interface between the modules and the auxiliaries. In general, the presence of multiple parallel service ducts beneath a false floor (above an interior deck), sets both the module height, and from the previously described aspect ratio consideration, in turn sets the width and length of a single-stack module. If multiple stacks are used in each module the position of the parallel service ducts must correspond to the underside of each stack, or a subsidiary manifold to feed the services to the multiple stacks must be placed between the bottom of each stack and the floor of the module. This would shorten the stack height and reduce the stack voltage, which is not desirable. The auxiliaries, such as the air handling system and the cooling and waste heat recovery system, can be placed in separate compartments and serve many modules at a time.

17. Module accessibility, removal

As a further refinement on the above described process, the means to physically translate individual fuel cell power plant modules horizontally from their operating locations onboard becomes an important tradeoff and ship interface issue. Horizontal movement approaches plus vertical lift through access openings or hatches to and from the 'fuel cell compartment' (which is analogous to the former 'engine room'), must be provided to facilitate the rapid changeout feature. The modules must not be too big for the hatch openings or too heavy for the lifting systems.

18. Electrical connections, controls

High power electric propulsion motors will require voltages of 4000 to 6000 volts d.c. If a typical fuel cell stack height in an available 8 foot high inter-deck space occupies 6 feet of height the stack voltage will then be in the order of 150 V. A 6000 V. d.c. electrical propulsion motor bus would thus require 40 of these stacks electrically connected in series to achieve the needed voltage. On the other hand, a lower power small ship might have a 600 V electrical bus voltage, which would only require 4 such stacks connected in series. The particular application will set the number and type of interconnections needed, plus the alternative connection possibilities in case of combatant battle damage or other outages of individual fuel cell modules.

The control system to keep the multiple fuel cell modules performing over a wide range of power levels must also be considered with care. Because fuel cells are d.c. output devices the use of d.c. power throughout the ship for auxiliaries is entirely appropriate. In the second world war time period and thereafter, many naval, maritime and commercial ships had 240 V d.c. auxiliaries as the standard.

This completes the surface ship process selection steps. The next two sections deal with the submarine application of fuel cells.

The Chief of Naval Operations, Admiral C. A. H. Trost, said in a September 26, 1988 Keynote Address at a Navy Research and Development Symposium, that "... I am declaring that integrated electric drive, with its associated cluster of technologies, *will be* the method of propulsion for the next class of surface battle force combatants, and I am directing all the major Navy organizations involved in these efforts to concentrate their energies toward that objective." [3]. Fuel cell power can be a part of that process.

19. Oxygen consumption rate and source

The last two steps in the process selection, No. 19 and No. 20, deal with the submarine application of fuel cells. Just as the fuel rate is important, as was discussed in Section 13 above, the undersea application of fuel cell propulsion power for submarines requres that the oxidant be carried along with the fuel. Because shallow submarines are generally considered to be volume-limited designs (except in the case of submarine tankers), the interior volume allocated to the fuel plus the oxidant will define the submerged endurance and hence the range of the submarine at a given speed. The lower the fuel rate is kept, the lower the oxygen rate will be as well. Thus the highest energy conversion efficiency power plant should be chosen. Because the molten carbonate type was analytically derived as the best approach for surface ship power it is also the best for submarine propulsion power [4]. When operated on pure oxygen the efficiency rises a further 5% compared to operating on air. Liquid oxygen (O_2 (l)) is the most compact or volume-efficient means to carry the needed oxidant. Many other issues arise in the fuel cell propelled submarine case, but that is the subject of another paper.

20. Pressurization, pressure compensation

Another means to improve the energy conversion efficiency, reduce the fuel rate and reduce the oxygen rate (also in kg/kW h), is to pressurize the fuel cell power plant. Because submarines operate under the pressure of the sea at cruise depths, the naturally available ambient sea pressure permits the fuel cell system to be pressure balanced to the ambient sea pressure. For example, a 16 atmosphere pressure, equivalent to roughly a 500 ft depth, will increase the voltage, and hence the power output, by a further 28% compared to the system's power output with 1 atmosphere of oxygen pressure. Thus, a 5 MW propulsion power plant operated near to the surface would produce 6.4 MW of power at a 500 foot depth of submarine operation. The submarine could thus travel faster at-depth, or the fuel and oxygen could be made to last longer for the same speed of advance.

This completes the Fuel Cell Process Selection discussion. The analytical process selection steps should be undertaken in an ordered sequence. The particular sequence of steps was given considerable thought and the technique is now offered to the reader as a proven system engineering approach.

References

1 Fuel cell propulsion technology assessment, ship impact analysis for combat ship-building, *Tech. Rep.*, John J. McMullen Associates Inc. and Arctic Energies Ltd. for the U.S. Naval Sea Systems Command, Washington, DC, Sept. 1987.
2 Marine applications for fuel cell technology, a technical memorandum, *Tech. Rep. OTA-TM-0-37*, Office of Technology Assessment, U.S. Congress, Feb. 1986; U.S. Government Printing Officie, Washington, DC, Library of Congress Card Number 85-600642.
3 Admiral C. A. H. Trost, Chief of Naval Operations, The cresting wave, Keynote address, *U.S. Navy League R&D Symp.*, Washington, DC, U.S.A., Sept. 26, 1988.
4 W. H. Kumm, Arctic Energies Ltd., Prospects for commercial submarine ships, *Shipbuilding Technol. Int.*, Feb. 1988, London, U.K.

Journal of Power Sources, 29 (1990) 181 - 192   181

POSSIBLE FUEL CELL APPLICATIONS FOR SHIPS AND SUBMARINES*

V. W. ADAMS

DGME-Procurement Executive, Ministry of Defence, ME 511, Rm 72, B Block, Foxhill, Bath BA1 5AB (U.K.)

Introduction

Fuel cells chemically convert fuels into direct current electrical energy and unlike heat engines, are not limited by the Carnot cycle. Although still an emerging technology, small and high reliability fuel cell plants have been built for space applications and large systems as demonstrators for commercial power generation [1]. In these applications automatic and reliable operation of fuel cells has been demonstrated.

Military applications have included small land based systems using phosphoric acid electrolyte fuel cells [2, 3] and a German development using alkaline electrolyte fuel cells in a submarine [4, 5].

For the operation of a fuel cell system, fuels generally have to be converted to usable hydrogen which is combined with oxygen in the fuel cell to produce water and electric power through a load. The following summarises benefits which might accrue from applying fuel cell systems to surface ships and submarines, and is based on earlier papers [6, 7].

Advantages and disadvantages

The advantages of replacing diesel-generators with fuel cell systems are seen to be:

(a) high efficiency (50% to 65% compared with 25% to 35% for a diesel-generator) resulting in increased endurance

(b) lower noise output (no moving parts except for pumps for fuel/air supplies) and IR signature

(c) lower running and maintenance costs with a significant increase in the mean time before module replacement compared to diesel electric generators

(d) savings in weight

---

*This paper expresses the views of the author which do not necessarily represent those of his department.

                Elsevier Sequoia/Printed in The Netherlands

√(e) direct generation of d.c. for supply to low/medium speed electric propulsion motors

(f) the ability to disperse fuel cell generators where required throughout the vessel

(g) a preliminary estimate indicates that there is no increase in through-life costs when compared with diesel-generatores.

Operational advantages in terms of broad operational requirements are therefore seen to be:

(a) reduced fuel consumption because of increased efficiency

(b) lower noise signature output because of fewer moving parts and a lower IR signature because of lower exhaust temperature and less waste heat

(c) increased endurance because of lower fuel consumption

(d) lower manning requirement because of installation of automatically controlled system and reduced maintenance

(e) lower support costs (running and maintenance), because of installation of modular system

(f) increased availability because of increased reliability

(g) increased battle survivability because of dispersion of fuel cell generation throughout the ship

Disavantages of fuel cell systems are seen to be:

(a) new development and higher procurement costs than diesel-generators (approx. three times for procurement)

(b) the need for d.c. to a.c. conversion for auxiliaries

(c) in the case of submarines, the need to dispose discreetly of waste products such as water, heat and (for reformed fuels) carbon dioxide,

(d) the need to demonstrate the ability to reform diesel fuel into gases suitable for use with fuel cells

(e) possible difficulties in the operation of reformers and fuel cells; depending upon the design, these may both need to be modularised and one or more modules run at full power to achieve variation in the output.

Submarine systems

Submarine application requires the vessel to carry both fuel and oxidant which can readily be used by the fuel cell system or converted to hydrogen and oxygen before being used.

A submarine system based on metal hydride and liquid oxygen stores has been developed in the F.R.G. [4] and a prototype auxiliary system capable of generating 100 kW fitted into an extended type 201 (U1) for the Federal German Navy [5].

Fuel cell technology and performance is very dependent on the type of fuel cell as well as the type of fuel and oxidant used. Studies [6] have indicated that cells using either proton exchange membrane (PEM) electrolyte or alkaline electrolyte will best meet a submarine's needs primarily due to high power density, low temperature operation and fast start-up times, PEM being the preferred option. Some developments in fuel cells [8, 9] confirm this. Other studies [10, 11] show a different conclusion, *i.e.* that molten carbonate electrolyte cells would be advantageous.

The choice of fuel for underwater vehicle storage is wide; for example, hydrogen, alcohols, hydrocarbons, hydrazine or ammonia could be used. However, some are very reactive or expensive and the final selection depends on such criteria as reactivity and method of storage, or whether reforming is possible to produce a fuel useable in fuel cells.

Since submarines spend only a small amount of their running time at full speed it is assumed that it would not be cost effective to install large power fuel cells for this purpose. Thus, batteries would be retained for high speeds and, depending upon the installed fuel cell power, diesel-generators may be needed for battery recharging.

In order to obtain maximum efficiency the choice of oxidant is limited to gaseous or liquid oxygen, or chemicals such as peroxides or heavy metal oxides which can produce oxygen by dissociation. Table 1 lists data for oxidant and fuel stores which are considered to be the most practicable and cost effective.

TABLE 1

Fuel and oxidant data relative to overall storage space available

Fuel	Oxidant	Possible type of stowage	Store	
			kg O_2/m^3	kg H_2/m^3
Diesel		tanks		280
Methanol		outboard flexible containers or tanks		100
Hydrogen		solid metal hydride		10 to 25
	HTP	outboard flexible containers	440	
	$O_2(l)$	cryogenic tanks	100 to 350 (depends on pressure)	

To compare the various combinations of fuel and oxidant Fig. 1 shows the benefit, in comparative terms only, to the submerged endurance that might be obtained for a typical conventional (non-nuclear) patrol submarine of two to three thousand tonnes displacement. The variation in endurance at patrol speeds against overall stored volume of the various combinations of fuel and oxidant is illustrated. Typical endurances for a lead/acid battery

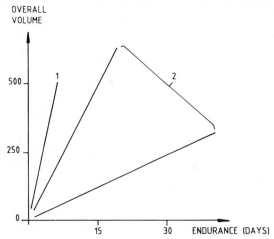

Fig. 1. Variations of volume/endurance dependent on fuel + oxygen combination: 1, lead acid battery; 2, fuel cell plant.

are also illustrated for comparison. By a suitable choice of fuel and oxidant a significant improvement in underwater endurance is possible with fuel cell systems.

Surface ship systems

Surface ship applications are even more favorable than in submarines because on oxidant for the fuel cell does not have to be carried and the disposal of waste products such as carbon dioxide and water does not present a problem.

It is considered to be impracticable to use fuel cell systems for the generation of sufficient power to drive large ships at their maximum speed. Thus, in these cases, application is confined to replacing on board diesel-generators used for power generation and low to medium electric ship drives, whilst the existing prime movers, such as gas turbines, are retained for the higher speeds. Hence, for surface ship application it would be highly desirable to employ fuel cells which can readily operate on diesel oil re-formate, ideally assumed to be a mixture of hydrogen and carbon dioxide, as a fuel and air as an oxidant. This would require the use of onboard diesel reformers.

Several types of fuel cell could be developed for surface ship application, but considerations here will be mainly restricted to those which are furthest advanced, can readily use diesel oil reformate as a fuel and air as an oxidant. Marine fuel such as diesel oil would need to be reformed into a reformate consisting ideally of a mixture of hydrogen and carbon dioxide which could be used directly in the fuel cell stacks. Other suitable types not yet fully developed will be mentioned briefly.

The above criteria limit us in the short term to:

(a) phosphoric acid fuel cells (PAFC), which can operate from re-formed fuel and are carbon monoxide tolerant

(b) solid polymer fuel cells (SPFC), also known as proton exchange membrane fuel cells (PEMFC), preferred due to their higher power density and fast start-up times, except that they are poisoned by carbon monoxide.

Other longer term development possibilities [1, 10] are molten carbon-ate fuel cells (MCFC) or solid oxide fuel cells (SOFC) and these offer the possibility of using integrated reformers because of their higher temperature of operation. The PAFC, SPFC and MCFC types would require the diesel fuel reformate to be desulphurised, whereas SOFC may be more tolerant to sulphur.

PAFC have been widely tested in land based demonstrators varying in power output from 40 kW to 4.5 MW and using reformed natural gas as a fuel [1]. However, PAFC are no longer receiving large funding for utility demonstration. SPFC with power output up to 5 kW have been built and larger cell stacks are being developed. Solid polymer electrolyte technology is also employed for electrolysers.

MCFC power plants are now receiving major electric utility develop-ment and demonstration funding in the U.S.A., Japan and Europe, leading to multimegawatt power level plants in the mid 1990s.

When considering the most suitable types of fuel cell, existing data are only available for PAFC and small SPFC. Data on 20 kW MCFC laboratory prototypes are also available. Figure 2 gives estimated variations of fuel cell system sizes, including reformers, compared with typical diesel-generators. Reformer parameters are more difficult to estimate than those for the fuel cell stacks and in Fig. 2, are based on land demonstrator units built for natural gas. Figure 3 gives estimated weights for fuel cell stacks only, com-pared with diesel-generators.

Estimated diesel oil consumptions for fuel cells are shown in Fig. 4 compared with diesel-generators. The data for fuel cells is based on the steam reforming of diesel oil into carbon monoxide and hydrogen followed by a water gas shift reaction to produce further hydrogen and carbon dioxide with an overall 90% conversion efficiency. Dodecane ($C_{12}H_{26}$) is assumed as a suitable model for diesel oil, but the choice of other hydro-carbons as a model would not change the results significantly.

In order to give some indication of fuel cell system module sizes, estimated outlines of cell stacks and reformers for surface ship application, based on known commercial developments and proposals, are shown in Fig. 5, together with typical outlines of diesel-generators. A possible schematic layout for a 1.4 MW SPFC system is shown in Fig. 6 compared on the same scale with a typical surface ship 1.3 MW diesel-generator in its housing. A volume for a 1.4 MW MCFC system based on published data [11] is also given for comparison.

It is not possible at this stage to compare complete installations of comparable power output.

VOLUME (m³) (POWER PLANT AND REFORMER)

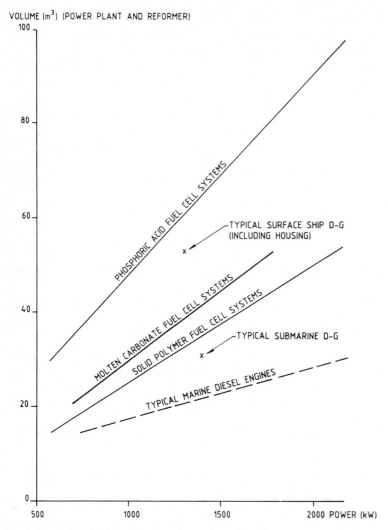

Fig. 2. Estimated volumes.

System engineering

Individual fuel cells must be combined into stacks and it has been identified that careful attention needs to be paid to the control of fuel, waste products and power management for any system to be successful. Fuel cells are readily assembled into stacks and stacks into modules and this has been successfully achieved for PAFC systems up to 4.5 MW [1]; technical risk is predominantly associated with system management. The risk involved in scaling up SPFC from the present 5 kW to larger modules for use in practicable systems is therefore assessed as low. A 50 to 100 kW

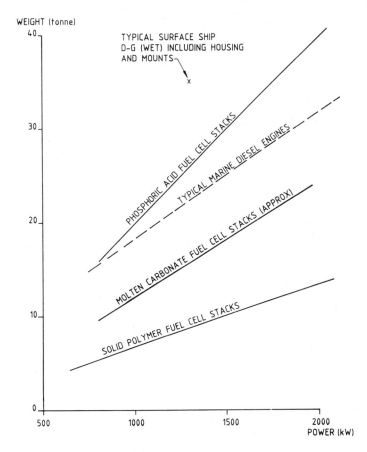

Fig. 3. Estimated weights.

module is the minimum thought to be suitable for building up into systems in excess of 1 MW.

In order to power surface ship auxiliary systems it would be necessary either to convert these to d.c. or, as is more likely, to provide d.c. to a.c. conversion equipment. Solid state d.c. to a.c. conversion equipment has been successfully demonstrated with fuel cell systems built as demonstrators for commercial power generation [1] and no problems are envisaged.

For submarine operation where a battery installation is retained it would be necessary to install a d.c. to d.c. power conditioner if the batteries are to remain connected during fuel cell operation.

Liquid fuels such as alcohols and hydrocarbons may be reformed to produce a reformate gas consisting ideally of hydrogen and carbon dioxide. This reformate may be fed to the fuel cells where the hydrogen combines with oxygen to form water and usable power is produced. In practice the fuel process involves steam reforming to produce carbon monoxide and hydrogen followed by a water gas shift reaction (using steam) to produce further hydrogen and carbon dioxide, *e.g.*

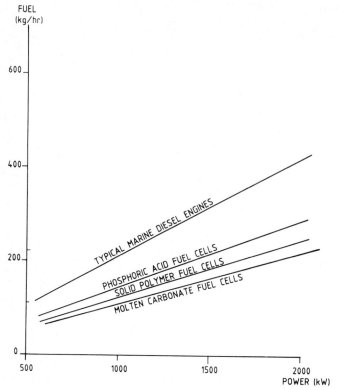

Fig. 4. Estimated fuel consumptions.

(a) $CH_3OH = CO + 2H_2$ and $CO + H_2O = CO_2 + H_2$ with an overall reaction of $CH_3OH + H_2O = CO_2 + 3H_2$ for methanol

(b) $C_nH_{2n+2} + nH_2O = (2n + 1)H_2 + nCO$ and $nCO + nH_2O = nCO_2 + nH_2$ with an overall reaction of $C_nH_{2n+2} + 2nH_2O = (3n + 1)H_2 + nCO_2$ for a hydrocarbon such as diesel fuel.

Steam reforming involves the use of steam and hence feed water. However, each mole of hydrogen produced from the fuel reacts with oxygen in the fuel cells to produce a mole of water. Thus, if the water requirement per mole of hydrogen produced is less than unity then the fuel cells can provide water both for reforming and other purposes.

Methanol requires less water than other candidate fuels with more water being required with increase in carbon number: based on stoichiometric reactions, water needed/mole hydrogen produced is 0.33 for methanol and 0.65 for $C_{12}H_{26}$ (an assumed model for diesel oil). In practice however, except for methanol, it is necessary to conduct the reaction in an excess of steam in order to suppress the deposition of carbon onto the catalyst used in the reforming process. (Methanol is unusual in that no excess steam is required because it readily thermally decomposes into carbon monoxide and hydrogen which does not lead to carbon deposition.) Thus, unless the

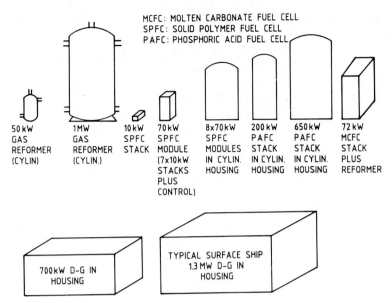

Fig. 5. Estimated overall sizes to same scale.

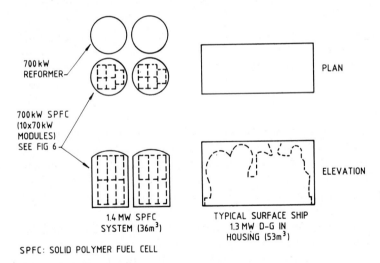

Fig. 6. Possible schematic layouts to same scale.

excess is removed downstream more will be required than is produced by the fuel cells.

Problems associated with diesel oil are the possible, but unlikely, presence of additives in readily available products that could poison the catalysts used in the reformer and the presence of sulphur.

Desulphurisation of the diesel oil is necessary to avoid poisoning the fuel cell catalyst and this possibly could be achieved after the steam reforming reaction by hydrogenating the sulphur compounds using some of the product hydrogen, thus removing the sulphur as hydrogen sulphide. The only candidate for the direct reforming of diesel oil (internally to the fuel cell system) without desulphurisation is the solid oxide type of fuel cell, although SOFC are not as sulphur tolerant as a diesel engine.

The U.S. DOD have funded work on reformers for diesel oil (not a marine application) [12]. This programme, for the USAF, included the demonstration of processes (including desulphurisation) capable of converting diesel fuel to fuel cell quality hydrogen for low power plants (less than 50 kW). Three demonstrators were funded in 1985 (Energy Research Corp. (ERC), International Fuel Cells (IFC) and R. M. Parsons Co. (RMP)) and run, although the IFC version failed after 20 h. The ERC plant employed desulphurisation followed by steam reforming and completed 400 h of satisfactory performance. The RMP plant used steam reforming followed by autothermal reforming and sulphur removal; this demonstration was inconclusive due to 'mechanical problems' but the process was considered to be a viable contender [12].

Compact gas reformers for PAFC are normally of tubular construction but these have the disadvantage that they are difficult to moderate in response to load changes. The reformer tubes are more liable to cause incomplete combustion (and hence carbon monoxide in the reformate gas) the smaller the size becomes, abnormal temperatures can occur in them in the event of load changes and the catalyst can become broken and choked up by lodgement.

More recently [13] a new plate type design of reformer under development in Japan is aimed at the improvement of reformer response to load changes and integration with high temperature fuel cells such as the MCFC type. The heat transmission and reaction mechanisms, and control of the gas temperature have been successfully demonstrated in a catalyst-filled plate reformer. It would be advantageous if this type of construction could be applied to diesel oil reformers.

Thus, there is some technical risk involved with the reforming function for diesel oil. Additionally, some reforming processes cannot be readily moderated to provide fuel to the cells for the generation of intermediate power levels. However, it is envisaged that, like the fuel cell stacks, reformers could also be modularised to provide fuel to each 50 to 100 kW, or larger, fuel cell module. Carbon monoxide is difficult to remove, particularly when the reformer is required to operate over a wide range of power levels from idle to full power with rapid throttle response. New designs of reformer based on a plate type of construction [13], may alleviate this problem.

Estimated costs

In order to assess the relative through-life costs the following assumptions have been made:

(a) Support costs for fuel cell systems would be the same as those for diesel-generators (but in practice are almost certain to be less)

(b) Capital cost of a 1 MW fuel cell system is £2M

(c) Capital cost of a 750 kW diesel-generator is £500K

(d) Fuel costs over a 25 year ship life are based on a 30% usage and the specific fuel consumptions in Fig. 4.

Based on the above the through-life costs for a 1 MW output are approximately £5.6M for both diesel-generator and fuel cell systems.

Conclusions

In terms of power to weight ratio, fuel consumption and operational advantages, projected solid polymer electrolyte and molten carbonate electrolyte fuel cell systems are potentially more than competitive when compared with diesel-generators.

Although a detailed through-life costing exercise needs to be carried out, indications are that the total through-life costs for a fuel cell system would not exceed that for a diesel-generator fit. Against this however, it is necessary to offset the unknown development costs for marine fuel cell systems, including both cell stacks and diesel oil reformers.

The potential advantages that are seen to be offered by fuel cell systems, particularly low noise signature and reliable automatic operation are considered to outweigh the disadvantages. Areas of technical risk in developing fuel cells for surface ships are the problems associated with the steam reforming of diesel oil and overall system management.

In the longer term, molten carbonate fuel cells offer the possibility of incorporating an integrated diesel oil reformer. Solid oxide fuel cells, not yet beyond the early development stage, offer the possibility of directly reforming moderately sulphur-containing diesel oil internally, *i.e.* without total desulphurisation.

References

1 *Abstr., Fuel Cell Seminars, U.S.A., 1985 - 1988*, The National Fuel Cell Coordinating Group, U.S.A.

2 S. Abens, P. Marchetti and M. Lambrech, 1.5 kW indirect methanol fuel cell, *Abstr. 1977 National Fuel Cell Seminar*, The National Fuel Cell Coordinating Group, U.S.A.

3 Unpublished work, Johnson Matthey Group, U.K.

4 K. Knaak, Fuel cell propulsion of submarines, *Maritime Defence*, 5/86 (May) (1986) 122 - 125.

5 R. Corlett, Air-independent German submarine power system development, *Maritime Defense*, (Sept.) (1988) 333 - 335.

6 V. W. Adams, Possible future propulsion systems, *Proc. Underwater Defence Technology Conf., London, U.K., Oct. 26 - 28, 1988*, Microwave Exhibitions and Publishers Ltd., pp. 537 - 543.

7 V. W. Adams, Fuel cell applications for surface ships, *Institute of Marine Engineers Centenary Conf., Royal Naval Engineering College, Manadon, Sept. 6 - 7, 1989,* to be published.

8 P. G. Patil, J. R. Huff and H. S. Murray, Fuel cell and fuel cell/battery power sources for vehicles, *Abstr. 1986 Fuel Cell Seminar, Tucson, AZ, U.S.A., Oct. 26 - 29, 1986,* The National Fuel Cell Coordinating Group, U.S.A., pp. 325.

9 Unpublished work, Ballard Technologies Corp., Vancouver, Canada.

10 W. Kumm, Marine fuel cell process selection applied to large surface ship fuel cell propulsion, *Fuel Cell Seminar, Long Beach, CA, U.S.A., Oct. 23 - 26, 1988,* The National Fuel Cell Coordinating Group, U.S.A., pp. 271 - 272.

11 W. Kumm, *Advanced Non-nuclear Submarine Propulsion: The Tradeoff Issues, 1989,* Arctic Energies Ltd, Severna Park, MD, U.S.A., March 1989.

12 W. G. Taschek, M. Turner and J. Fellner, Air force remote site fuel cell development programme, *Abstr. 1986 Fuel Cell Seminar, Tucson, AZ, U.S.A., Oct. 26 - 29, 1986,* The National Fuel Cell Coordinating Group, U.S.A., pp. 342 - 345.

13 IHI's MCFC system nearing completion, *Ishikawajima-Harima Heavy Industries Co. Ltd. (IHI) Bulletin, 22* (238) (Dec.) (1988) 4.

Journal of Power Sources, 29 (1990) 193 - 200

THE FUEL CELL IN SPACE: YESTERDAY, TODAY AND TOMORROW

MARVIN WARSHAY* and PAUL R. PROKOPIUS

Electrochemical Technology Branch, National Aeronautics Space Administration, Lewis Research Center, 21000 Brook Park Road, Cleveland, OH 44122 (U.S.A.)

Introduction

The first practical fuel cell resulted from work begun in the U.K. in 1932 by F. T. Bacon. Eventually a 5 kW hydrogen–oxygen, alkaline electrolyte system developed by Bacon demonstrated its capability by powering a welding machine, a circular saw and a two-ton forklift truck. With these and other demonstrations of the applications of this 'new' power device, the fuel cell had finally apparently emerged from the laboratory. However, it was the worldwide attention to NASA space missions that introduced 'fuel cell' to the vocabulary of millions of people. Ironically, it has probably been the announcement, during space flights, of real or suspected fuel cell malfunctions, rather than the usual smooth performance of the fuel cells in space, that has given fuel cells their wide recognition. (The aborted Apollo 13 flight was a case in point. A prelaunch malfunction of an oxygen feed control component — not the proclaimed fuel cell problem — was the real cause of the near disaster that attracted the attention of many millions of people.)

Past

In the early years of U.S. space flight, the fuel cell was selected over other competing power systems because of its greater promise to meet the on-board power requirements of planned NASA extended duration manned missions. In addition to satisfying the power, efficiency, weight, life, reliability, safety, mission flexibility, development maturity, etc. requirements, the fuel cell offered a number of special advantages over competing power systems. Noteworthy among these advantages was the ability of the hydrogen–oxygen fuel cell to supply potable water (the product of the electrochemical reaction) for crew consumption and for cabin air humidification.

What emerged as a result of the NASA selection of the fuel cell was an almost explosive growth in fuel cell research and development (primarily sponsored by NASA and other U.S. government organizations) in industries, universities and government laboratories.

---

*Author to whom correspondence should be addressed.

0378-7753/90/$3.50

For the Gemini earth-orbiting mission (1962 - 1965) fuel cells were successful in supplying power in a reliable manner. The General Electric (GE) fuel cells that were used for seven flights of that mission utilized solid polymer electrolytes (called ion-exchange membrane (IEM) at that time) consisting of a cationic membrane of sulfonated polystyrene resin. This type of electrolyte had mobile H^+ ions in well-defined electrolyte boundaries. The advantage of the obvious ease of electrolyte containment was offset by the ohmic resistance of the membrane, which contributed to the lower performance (voltage efficiency) of the IEM than of alkaline fuel cell systems such as that used for the Apollo missions that followed. Making the membrane thin minimized, as much as possible, the effect of high ohmic resistance. (In the 1980s there has been considerable improvement in the performance of this concept, now called the proton exchange membrane, or PEM. This is discussed in the 'Future' section.)

The Gemini 1 kW powerplant consisted of three stacks of 32 cells. The heat generated by the fuel cell stack was removed by a circulating coolant. Two 1 kW power plants were on board to handle maximum load requirements. The average power that was produced on the Gemini flights was 620 W. The hydrogen and oxygen reactants were stored in their cryogenic states. The nominal hydrogen supply pressure was 1.7 psi above water pressure, that of the oxygen 0.5 psi above the hydrogen pressure. The cell operating temperature was 70 °F. The anode and cathode consisted of titanium screens upon which Pt unsupported catalysts with ptfe were deposited. The catalyst loading was 28 mg Pt/cm^2/electrode. This IEM fuel cell technology was subsequently (1967) used for the Biosatellite spacecraft. An important change in the IEM fuel cell technology for this application was the use of a new membrane, namely the perfluorosulfonic acid ionomer. The membrane called Nafion (registered trademark) was developed by DuPont. These types of cationic membranes became the standard for this type of fuel cell, which continues to this day.

A special problem of the Gemini IEM fuel cell was its sensitivity to membrane water content. With insufficient water the membrane would dry out and often crack. On the other hand, the membrane could not hold too much water. A flooded electrode was often the result. Both extremes would result in a severe performance loss. To avoid the problem of excess water, the Gemini fuel cell design utilized wicks to carry excess water to a ceramic porous separator where the water was separated from the oxygen and sent to an accumulator for storage.

The fuel cell technology that went to the moon was not based upon the Gemini IEM fuel cell of the 1960s, but rather upon the Bacon cell that preceded the GE IEM fuel cell work. Through the British National Research and Development Council and Leesona-Moos Laboratories, Pratt and Whitney acquired the patent to Bacon's fuel cell technology in 1959 and applied the technology to the NASA Apollo mission. However, for space use the heavy, high pressure Bacon cell was not directly suitable. For the Apollo fuel cell, the pressure was lowered from 600 to 50 psi. To prevent the KOH from boiling at 205 °C, the KOH concentration was increased from 30 to 75%. But, at ambient temperature 75% KOH is solid. However, this proved not to be a significant problem. Finally, the temperature was raised to 260 °C to recover the performance lost by the pressure reduction. The Apollo fuel cell included Bacon's double-porosity layer nickel electrodes designed to maintain the gas–electrolyte interface at the boundary between the pore size regions. The anode was porous nickel while the cathode was lithiated, oxidized porous nickel. Because of the high temperature (maximum) of 260 °C a highly active catalyst like Pt was not needed as in the case of the Gemini fuel cell which operated at 70 °F. At 260 °C and a current density of 150 A/ft^2 the voltage was 0.87 V per cell, while at its nominal operating temperature of 204 °C it produced 0.72 V at 150 A/ft^2. The performance of the Gemini fuel cell was lower.

The Apollo fuel cell 1.5 kW powerplant consisted of three modules connected electrically in parallel. Heat and water removal were by hydrogen circulation. A glycol–water secondary coolant loop was also employed. The power range was 563 to 1420 W. Peak power capability was 2295 W at 20.5 V. It weighed 220 pounds. The module rating was 400 h; but it ran 690 h without failing. The Apollo missions were from 1968 to 1972.

The KOH–H$_2$O electrolyte solution was pressurized to 53.5 psia while each reactant gas cavity was maintained at 63 psia. The operating pressure of the system and relative pressure differationals affected the fuel cell performance. The latter determined the location of the reactant–electrolyte interface.

Present

Bacon might not recognize the 'grandchild' of his alkaline fuel cell today, the Orbiter fuel cell. The high pressure, very heavy construction of Bacon's fuel cell was already gone in the Apollo fuel cell. In the Orbiter fuel cell, United Technologies Corp. (the new name for the Pratt and Whitney fuel cell organization) dropped the dual porosity electrodes. In the place of free electrolyte, the Orbiter electrolyte held the 32% KOH electrolyte in an asbestos matrix. Another change was the cell temperature, which was reduced to 93 °C. At this temperature an electrocatalyst was required to achieve a reasonable performance. The operating pressure is 60 psia. The electrodes consisted of gold plated Ni screens upon which a catalyst layer and ptfe were applied. The hydrophobic ptfe provided gas passages through the electrode. The catalyst loading on each electrode is 20 mg/cm^2 Au Pt alloy on the cathode and 10 mg/cm^2 Pt on the anode.

Heat generated by the fuel cell reaction is transferred to the fuel cell's coolant system. The coolant system, containing a fluorinated hydrocarbon dielectric liquid, transfers the heat through the Orbiter's heat exchangers to the freon coolant system.

The Space Shuttle Orbiter is equipped with three fuel cell powerplants supplying 12 kW at peak at 7 kW average power. Each powerplant weighs 250 pounds. The Orbiter's fuel cell powerplants are 50 pounds lighter and deliver up to eight times as much power as those of Apollo.

The fuel cell power plants are started approximately 8 h prior to launch, using ground-supplied hydrogen and oxygen reactants. Approximately seven minutes is required to bring the powerplants to full operating capacity. After start-up, the fuel cells share spacecraft electrical loads with ground power support. About three minutes prior to launch, the spacecraft automatically switches to onboard reactant supply and the fuel cells become the sole source of electrical power for the spacecraft for the duration of the mission. Approximately every eight hours during the mission each fuel cell powerplant is purged for two minutes to remove inert gases from the system.

Future

NASA's planning for the future exploration of the Solar System includes the establishment of manned outposts, as well as central base stations on the Moon and Mars. Supporting human expeditions to, and operations on, the surface of the Moon or Mars represents a substantial technology challenge for current and projected power system capabilities. The high levels of power associated with an operational base, somewhere in the hundreds to thousands of kilowatts, will require nuclear power systems. During the installation of these permanent nuclear systems, power systems based on solar energy hold the greatest promise for supplying needed power. These systems will also be required to augment and serve as back-up power sources for the permanent nuclear powered bases.

Because the solar-based surface power system must supply usable power continuously, that is during the day as well as the night, a regenerative system is required. During the daylight hours the power generation subsystem will recharge the energy storage subsystem and also supply power directly to the system's electrical loads. Thus, continuous power is supplied to the load; it is provided by the power generation subsystem during sun periods and from the energy storage subsystem during periods of darkness.

In a Lunar application, the period of darkness extends for two weeks, while a Mars application presents a more manageable 12-hour night. Both applications require very high energy density and reliable energy storage systems. The highest potential for successfully achieving surface power storage capabilities for these applications lies in the regenerative fuel cell (RFC) concept. The regenerative fuel cell system is depicted in Fig. 1. During the light portion of the orbit the photovoltaic solar arrays generate sufficient power to service the system electrical loads plus a water electrolysis unit. The amount of electrical energy required by the electrolysis unit is dictated by the amount of hydrogen and oxygen needed to generate power in a fuel cell which supplies the electrical loads during the dark portion of the orbit.

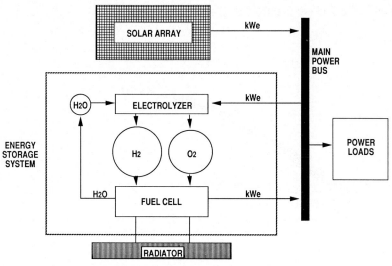

Fig. 1. Regenerative fuel cell system schematic.

In generating this power, water is produced by the fuel cell as a by-product of the electrochemical reaction. To complete the cycle, the by-product water is collected and stored for use in the electrolyzer during the succeeding orbit.

The mass and specific energy benefits to be realized by employing a regenerative fuel cell system are displayed in Fig. 2. Low system mass for a given power level is a central requirement for achieving acceptance of transportation costs to the Moon or Mars. Another requirement, even more challenging, is appreciable system lifetime without sacrificing performance even after an extended period of dormancy. Also a relatively high power level requirement of 25 kW is projected to support an initial surface outpost of four to six astronauts. To develop the technology base for a system which will meet these requirements, a program has been initiated as one of the elements of NASA's Project Pathfinder. This program was developed and is being managed by NASA's Lewis Research Center. It focuses on the technology areas of solar power generation, energy storage and electrical power management. Advancing these technologies and coupling their performance potentials with an advanced low mass, a reliable electrical power management subsystem can lead to surface power systems having a reliable life in excess of 20 000 hours with system specific powers of 3 W/kg for Lunar application and 8 W/kg for Martian applications. These projected specific powers represent substantial improvements over the state-of-the-art, up to a factor of 30. System mass reductions of this magnitude coupled to the expected factor of 10 increase in life, should enable extra-terrestrial surface missions where life and mass are the driving forces for success.

The energy storage element of the Pathfinder Surface Power Program is a 10-year effort culminating in the verification of a regenerative fuel cell system breadboard operating in a relevant environment. The near term, 5-year, Phase I effort, will provide the development and verification of the system

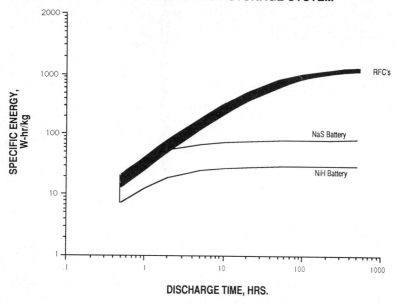

SPECIFIC ENERGY OF 25 kWe ENERGY STORAGE SYSTEM

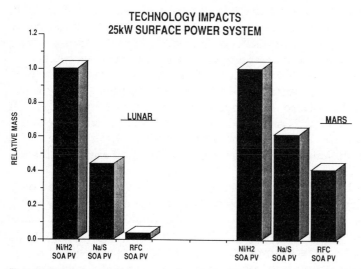

Fig. 2. Advantages of regenerative fuel cell energy storage *vs.* battery systems for long discharge applications.

critical components, those being the fuel cell and electrolyzer stacks. The second 5-year phase will focus on the development and verification of the complete RFC breadboard system

The two candidate fuel cell and electrolyzer technologies for the Path-finder system are the alkaline and proton exchange membrane (PEM). Because alkaline was the system of choice for both Apollo and the Space

Shuttle, the state-of-the-art of alkaline systems had been advanced considerably over that of the PEM technology. However, the recent technology efforts on fuel cells for transportation applications have advanced the PEM technology.

The major deficiency facing the alkaline technology in the Pathfinder application is the lack of long term catalyst layer stability, which translates into performance degradation with time. Unlike with PEM and other acid-type fuel cells, a stabilizing catalyst support has not been developed for the alkaline system. PEM, on the other hand, offers a stable, long life system but one whose efficiency has, until recently, been significantly lower than alkaline. Recent improvements in the conductivity of PEM membranes increase the probability that this technology could replace alkaline as the Pathfinder RFC baseline. At present, the weakness in the PEM technology stems from the fact that the membrane technology improvements are very recent and, therefore, the data base needed to justify commitment to this technology does not exist. Accordingly, a technology assessment has been undertaken to provide guidelines for selecting the technology to be carried into full development in the Pathfinder Program.

Since the late 1960s the U.S. Air Force has been supporting fuel cell technology development for future space applications requiring very high power densities for much shorter periods than for NASA missions. Figure 3 illustrates the steady progress over the years in alkaline fuel cell power density performance improvement. The work was carried out by UTC (this part of UTC is now called the International Fuel Cells Corp., IFC).

The solid oxide fuel cell (SOFC) in its monolith configuration has the potential for even higher power density performance than does the alkaline fuel cell system. However, the high power density alkaline fuel cell system is much further along in its development than is the SOFC for the Air Force Space applications. The government funding for the SOFC monolith concept has been directed at the NASP (National Aerospace Plane) application.

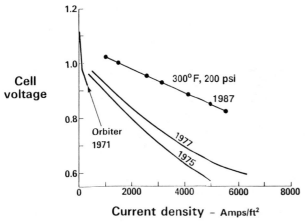

Fig. 3. Alkaline high power density performance.

Finally, the alkaline fuel cell system holds particular promise for the proposed National Space Transportation System (NSTS), sometimes referred to as the all-electric shuttle. Here the fuel cell is to supply both on-board power and high power density, short burst power for electrical control system actuators.

The European space programme also plans to use fuel cell systems to satisfy spacecraft power requirements. Hermes, the European manned reusable space plane will require 3 - 4 kW for low earth orbit missions. Its electrical system will utilize fuel cells as the primary power source and lithium primary batteries as a back-up/peak power supply (peak of 15 kW). For future European spacecraft high power requirements, European organizations have been studying RFC systems.

Conclusions

Figure 4 is a graphic depiction of the progress in space fuel cell power, as well as the hope for the future in particular applications. (However, this Figure does not depict the progress leading to the important future NASA space fuel cell application discussed in this paper, namely the RFC for Lunar and Mars surface power energy storage.) In terms of specific weight it illustrates the steady improvement from the past to the present, from the close to 200 lb/kW of the Apollo 1.5 kW powerplant to the 20 lb/kW of the Orbiter 12 kW fuel cell powerplant of today. Based on technology development both underway and planned, it forecasts meeting the goals of (1) about 1.5 lb/kW, in about 1993, for the 300 kW NSTS fuel cell powerplant, and (2) about 0.5 lb/kW for the very high power density, short duration applications at the beginning of the 21st century.

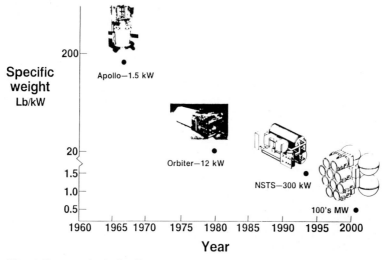

Fig. 4. Progress in fuel cell space power.

Journal of Power Sources, 29 (1990) 201 - 203 201

ALKALINE FUEL CELLS AT ELENCO

H. VAN DEN BROECK

Elenco NV, Gravenstraat 73 bis, B-2480 Dessel (Belgium)

Introduction

Elenco has been working on fuel cells since 1976. Its present capital is owned by DSM (Netherlands; 42.5%), SCK (Belgium; 42.5%), Euroventures Benelux (Belgium and the Netherlands; 14%) and the management.

Until 1985 work was exclusively on hydrogen–air; from then onwards both hydrogen–air and hydrogen–oxygen have been pursued.

General status of the technical development

In the 1976 - 1985 period the basic electrochemical hardware was developed. Figure 1 shows a 4.5 kW H_2–air stack. Further information can be found in ref. 1.

From 1984 onwards technical development has largely been focused on the fuel cell system as a whole, including all ancillary equipment needed. A general description of this equipment is also presented in ref. 1.

Fig. 1. 4.5 kW H_2–air fuel cell stack.

0378-7753/90/$3.50

Application areas and projects

At the present stage of development, it is felt that the main applications of the Elenco type alkaline cell are to be found in space, defense, electric traction and some specific stationary situations.

As the alkaline cell needs pure hydrogen, excluding substantial amounts of CO_2 and CO, applications must fit into scenarios where this fuel is or can be made available.

The main projects in which Elenco is involved are the following.

● Space: Elenco is part of an industrial team with the German companies Dornier and Siemens for the development of the fuel cell system for the ESA space plane Hermes

● Defense: a first series of six 1.2 kW fuel cell systems has been manufactured and is under assessment by several users

● Traction: after positive test results obtained with the Elenco fuel cell stacks in an electric van (Fig. 2), projects for the implementation of fuel cells in large commercial vehicles (such as city buses, refuse collecting trucks) are being set up.

First city bus project

The definition phase of a first city bus project has practically been concluded, and the start of the construction phase is imminent. Elenco's industrial partners for this project are Air Products Nederland, Holec and Den Oudsten, all of them companies based in the Netherlands. The prototype bus is to be tested first in the city of Amsterdam.

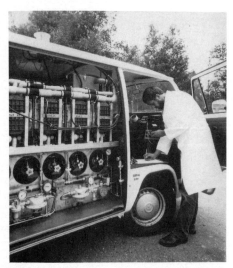

Fig. 2. Testing of fuel cell stacks in an electric van.

Air Products will take care of the hydrogen part in the project (storage on the bus and tanking facility in the bus garage); liquid hydrogen will be used.

Elenco will take care of the fuel cell system, and Holec of the electric traction components (the a.c. motor, the d.c./a.c. inverter and other electronics).

Finally, bus builder Den Oudsten will build the bus and will be in charge of obtaining all official authorizations.

The fuel cell size will be 70 kW and a mechanical energy accumulating system will be used for acceleration.

The design of the bus is such that its general specifications will be comparable to those of a diesel bus.

Reference

1 Status of Elenco's alkaline fuel cell technology, *Proc. IECEC Meeting, Philadelphia, PA, U.S.A., Aug. 10 - 13, 1987.*

Future Perspectives

Journal of Power Sources, 29 (1990) 207 - 221   207

EUROPEAN SPACE AGENCY FUEL CELL ACTIVITIES

F. BARON*

European Space Agency, Postbus 299 2200 AG, Noordwiek (The Netherlands)

Hermes activity

At the Ministerial conference in the Hague in November 1987, the development of the Hermes winged-space vehicle solution was endorsed and the first step of the development programme was approved.

This paper describes the progress status concerning the Hermes fuel cell development programme (HFCP).

Two fuel cell power plants represent the main power sources for Hermes which will consume an average of about 4.6 kW and an energy around 1220 kW h for a twelve day mission (including safety margin). Figure 1 shows a schematic of the power architecture.

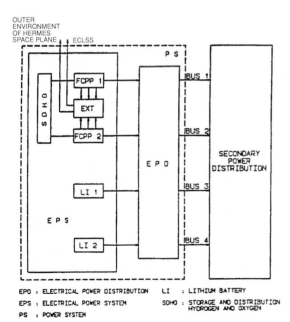

EPO : ELECTRICAL POWER DISTRIBUTION   LI : LITHIUM BATTERY
EPS : ELECTRICAL POWER SYSTEM   SOHO : STORAGE AND DISTRIBUTION
                                               HYDROGEN AND OXYGEN
PS : POWER SYSTEM

Fig. 1. Present architecture of the Hermes power system.

*Present address: ESA Toulouse, HERMES Engineering Division, 18 Avenue Edouard-Belin, 31055 Toulouse Cédex, France.

0378-7753/90/$3.50

TABLE 1

Main HFCP requirements

Power	2 - 6.5 kW
Voltage	75 - 115 V d.c.
Power/weight ratio	50 W/kg
Power/volume ratio	40 W/l
Water production at 4 kW	1.545 kg/h
Accumulated operation time	4000 h
Accumulated storage time	26000 h

The main present Hermes requirements are summarised in Table 1.

Under ESA funding, an industrial team led by Dornier, under direct supervision of Aerospatiale (prime contractor of the spaceplane) is performing the development task. Up till now, no European company has experience of fuel cells for space applications. Therefore, and because European experience in the field of fuel cells was gained in an alkaline medium, several different technologies based on an alkaline electrolyte and which can meet in principle the Hermes requirements, have been evaluated during a selection phase performed by Dornier in 1988.

The Hermes fuel cell power plant (HFCP) consists mainly of seven sections as follows.

- Stack
- Water separation device
- Cooling components
- Gas management elements
- Product water management
- Controller
- Harness, structure and containment

Although all the HFCP components are important, it was decided to concentrate efforts on the stack and the main peripheral components in a first step. Following a formal selection procedure prepared by Dornier and based on analytical and testing investigations and analysis, four basic fuel cell configurations have been evaluated.

Three companies have been pre-selected as potential suppliers and have been requested to demonstrate the maturity of their own technology and state-of-the-art. Elenco (B) for recirculation technology, Siemens (D) for recirculation and static technologies and Varta (D) for recirculation and static technologies.

The following summarises the features of the different technologies:

- Recirculation system (1). The circulating electrolyte (KOH) moves through the stack and is used to remove both the product water and the heat dissipated by the stack. A schematic of this configuration is described, see Fig. 2.

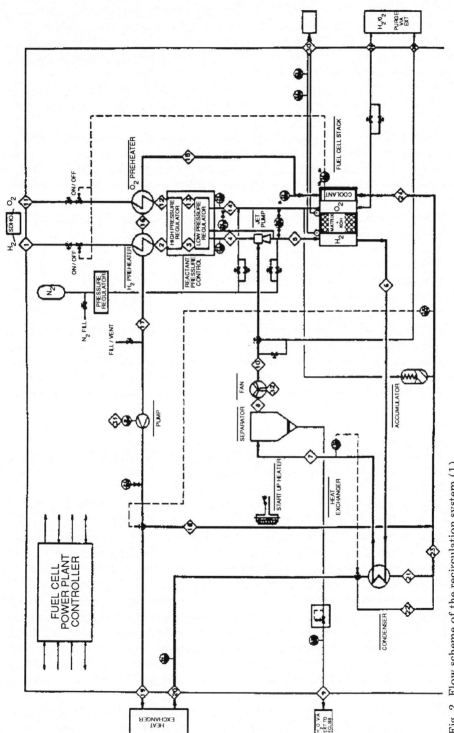

Fig. 2. Flow scheme of the recirculation system (1).

• Recirculation system (2). An alternative to the previous configuration where product water is now removed via the hydrogen gas loop, see Fig. 3.

• Static system. The electrolyte is retained in a matrix whereas a separate cooling loop removes the heat and the hydrogen loop removes the product water of the reaction as shown in Fig. 4.

• Static Eloflux system. The KOH is trapped inside the stack and an isotonic fluid cycle permits water removal and is used as cooling fluid (see (Fig. 5).

Several trade-offs related to the choice of sub-assembly dealing with the control of the HFCP (*e.g.* water management, gas management, thermal management) had to be investigated (still in progress) in order to optimise the scenarios/solutions with respect to the fulfilment of the Hermes requirements and the functionality of the overall power source. As a consequence, evaluation of non-electrochemical components has also been carried out with tests on ancillaries such as:

• Gas trap
• Jet pump, KOH pump
• Liquid/gas separator (hydrophilic membrane and cyclone)
• KOH regenerator
• Bubble separator (gas/liquid separation device)

All data collected from the different elements/components tested at Dornier (electrodes, stacks, peripherals/ancillaries) have formed the main input for the performance and functional engineering analysis of the system selection phase. With the help of a mathematical simulation model called SANFU, the modelling of system behaviour with respect to controlling aspect, evaluation of the operational functions, calculation of the overall performance data have been possible as well as the preliminary detection of critical areas.

Besides the 'raw technical data', other criteria have been taken into account for the selection. Table 2 describes the weighting factors which have been applied.

The compilation of the different results with the application of the criteria as listed in Table 2, showed that the static KOH system based on the Siemens technology was most advanced. After a critical analysis of weak points and safety aspects, it appeared that the static KOH technology proposed by Siemens has the best characteristics and presents the best growth potential regarding the space conditions/applications. Therefore, Dornier has recommended the development of this technology, a decision which has been supported by Aerospatiale. The customers (CNES and ESA) have requested Dornier to propose and form a strong and competent industrial team able to perform the next development phases.

At the time of writing this paper, important progress has been achieved and the necessary authorisations to proceed to the C1B phase should be given very soon.

The basic flow scheme configuration of the static KOH system is shown in Fig. 6 and the main functions can be summarised as follows.

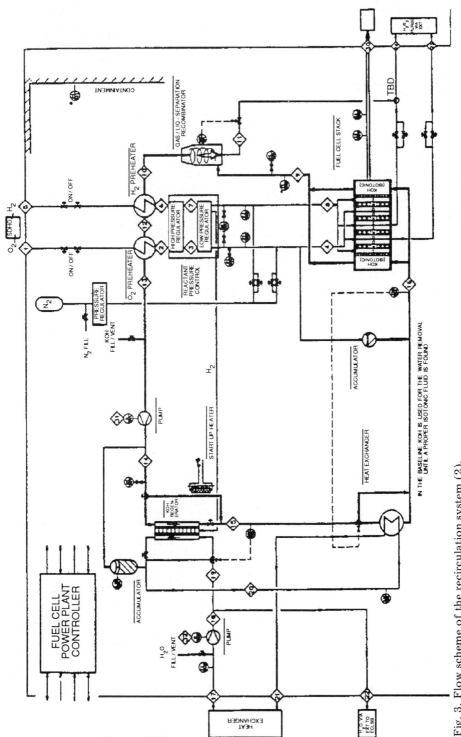

Fig. 3. Flow scheme of the recirculation system (2).

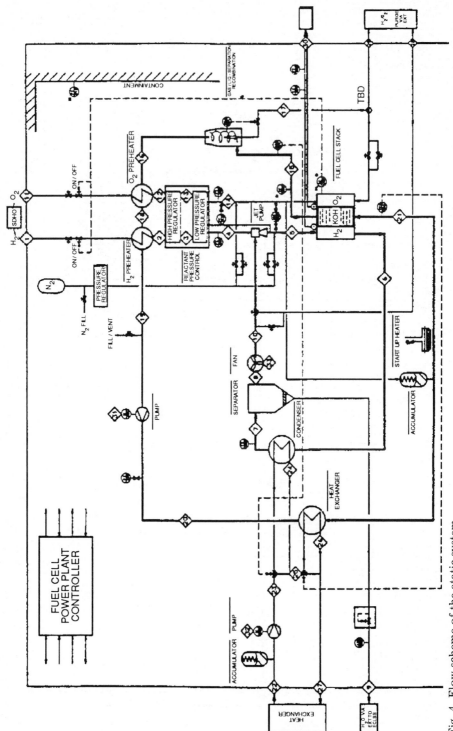

Fig. 4. Flow scheme of the atatic system.

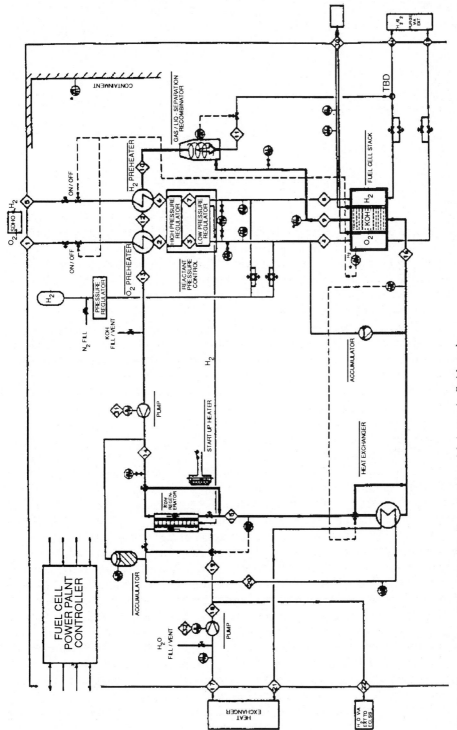

Fig. 5. Flow scheme of the static Eloflux system with isotonic fluid cycle.

TABLE 2

Selection criteria (%) with weighting factors

Safety	17
Performance	17
Reliability	15
Operational behaviour	15
Technological experience (including development risk)	10
Mass	10
Volume	7
European state-of-the-art	5
Cost	4

• The reactants coming from the gas distribution system (SDHO) are pre-heated above 0 °C and the pressure reduced from 65 (maximum) to 10 bar in a high pressure regulator device. Then, a low pressure regulator component finally reduces the reactant pressure in accordance with requirements and follows the pressure demand difference which is required for an optimum operation of the fuel cell stack.

• The hydrogen gas is transported with the help of a jet pump with an optional support of a fan in order to overcome the pressure drop across the peripherals where hydrogen is circulated. A significant excess of hydrogen is necessary to remove the product water. Humidified gas passes through a liquid/gas separator (either a membrane separator where water vapour diffuses or a pump separator combined with a condenser) where water is directed towards the product water management sub-equipment and the controlled and partially dried hydrogen is 're-injected' inside the H_2 compartments of the stack.

• A coolant fluid (e.g. water) is heated up when circulating inside the stack and is used to warm-up the reactants (with the pre-heaters) coming from the SDHO. Then a heat exchanger connected to the TCS (thermal control system) removes most of the heat dissipated by the fuel cell stack.

• A local controller will be in charge of all the functions of the fuel cells components: acquisition of data, sequence control and safety aspects. A multitude of sensors and actuators will be managed by the controller.

During phase C1B (ending in December 1990), the industrial development team led by Dornier and formed by Dornier, Siemens, Elenco and Drager is being requested to demonstrate the feasibility of the selected technology and to solve the main problems which are expected so far.

Considering the criticality of the fuel cells for space application, in particular for the limited time schedule of Hermes, it has been decided as a first priority that a full scale breadboard will be built and tested in order to perform a functional demonstration of the overall HFCP.

The main objectives being to have the first flight models (fully certified for the Hermes application) beginning of 1997, several development models

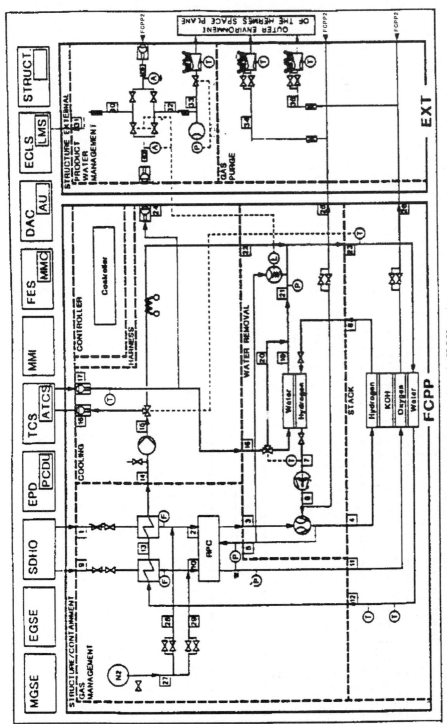

Fig. 6. Flow scheme of the technical baseline configuration of static KOH system.

216

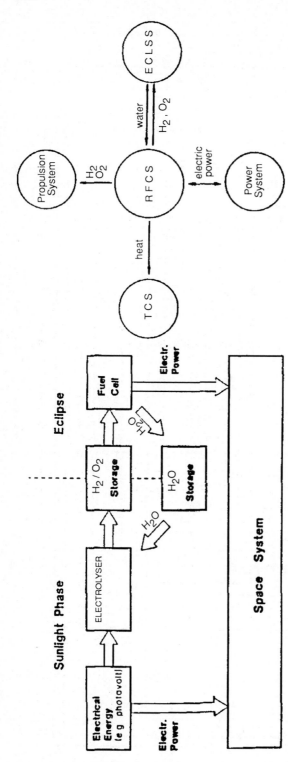

Fig. 7. Cyclic process of the RFCS and potential interfaces with other space systems.

TABLE 3

RFCS targets

Mean power	20 kW
Voltage level	120 V d.c.
Mission duration	5 years (30000 cycles)
Orbital replacement unit concept	
Reactant storage	30 bar (gas)
Overall mass	330 kg
Energy density	36 W h/kg

(engineering models and qualification models) will be tested within the C2/D phases.

RFCS activities

Since 1985, under the technology and research programme (TRP), ESA is funding preparatory activities related to the development of the RFCS as alternative to the conventional batteries (nickel–cadmium or nickel–hydrogen) commonly used as energy storage in combination with solar generators for European spacecraft and platforms. At power levels above 10 kW, the RFCS becomes attractive and lighter than other systems. Electrical power is provided by solar array during the sunlight phase to the electrolyser unit which regenerates from water, the hydrogen and oxygen reactants. These gas reactants are stored and supplied to the fuel cell which produces energy and water to the spacecraft during the eclipse phase.

The RFCS which works on a cyclic process mode could be combined with other space systems such as ECLS (where oxygen in excess can be provided to the life support, and hydrogen for the CO_2 reduction or propulsion system (supply of hydrogen and oxygen)). Figure 7 shows the operating principles of the RECS and its potential multi-functions with other space systems.

Two study phases have already been awarded to Dornier in order to review the feasibility of developing a European RFCS for space application, a third TRP phase is under negotiation (mid-1989 until end of 1991).

Although no short-term project has foreseen the use of RFCS, medium and long term ESA programmes for large low earth orbit platforms such as COLOMBUS-AOC (autonomous operational concept) are among the potential users. As a consequence, it is proposed to proceed later with the development of a 'breadboard' under project funding namely the European Manned Space Infrastructure (EMSI) budget.

Table 3 summarises the main requirements/targets of the RFCS.

During the first phase, four configurations were pre-selected.

- Static alkaline (KOH) electrolyte system

• Mixed system with KOH static fuel cell and PEM electrolyser
• Proton exchange membrane (PEM) system
• Recirculating alkaline (KOH) electrolyte system

During these two phases, a serious effort has been devoted to the study of alkaline and PEM potential electrolysers with European companies (CJBD (U.K.), Hydrogen System (Belgium), CGE (France)). Other tasks on main peripheral components such as the pumps, the gas and water storage tanks, the two-phase (G/L and L/G) separators have provided analytical data which have been used for system comparison with the help of a mathematical simulation model called SAREF.

A global comparison of the characteristics (performances, round trip efficiency etc.) has been performed by Dornier and its subcontractors during phase 2. Figures 8 to 11 summarize the results.

A review of advantages and disadvantages at system level of the different technologies/configurations has shown that the static KOH system and

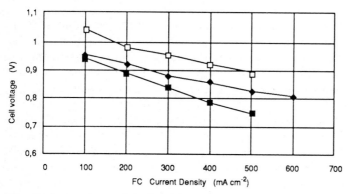

Fig. 8. Comparison of fuel cell system efficiencies, including thermal and current efficiency and auxiliary power demand. ■, Recirculation; □, static; ♦ PEM.

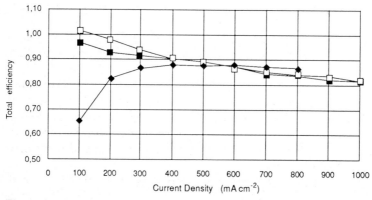

Fig. 9. Comparison of electrolyser system efficiencies, including thermal and current efficiency and auxiliary power. ■, Recirculation; □, static; ♦ PEM.

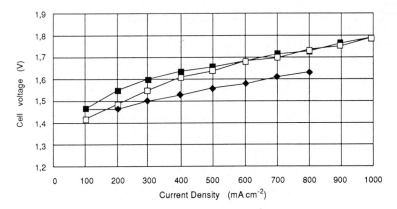

Fig. 10. Comparison of electrolyser performances (80 °C, 30 bar). ■, Recirculation; □, static; ♦, PEM.

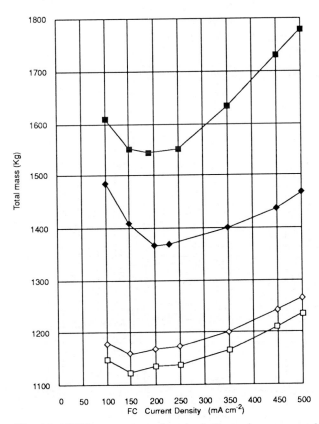

Fig. 11. RFCS mass comparison including solar arrays and radiators, power output 20 kW at 120 V, 0.6 h eclipse, 0.95 h sun. Optimum current densities of the electrolysers: static KOH 310 mA/cm², recirculating KOH 480 mA/cm², PEM 800 mA/cm². ■, Recirculation; □, static; ♦, PEM; ◊, mixture.

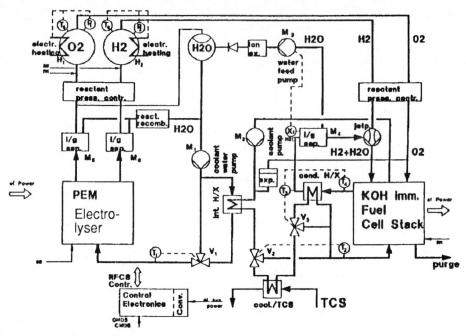

Fig. 12. 'Mixed' RFCS (static KOH fuel cell and PEM electrolyser).

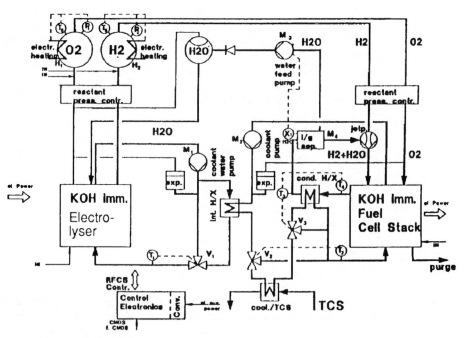

Fig. 13. Static KOH RFCS.

the mixed system stand out as most appropriate at this time for this specific space application.

Figure 12 describes the basic schematic configuration of the 'mixed' RFCS whereas Fig. 13 shows one of the complete static KOH RFCS. The main difference consists in the use of a PEM electrolyser instead of a static KOH electrolyser. The performances of these two different electrolysers are similar, the PEM electrolyser has the advantage at higher current densities (see Fig. 11). Nevertheless the main drawback of the PEM electrolyser is the relatively bad current efficiency due to the gas permeation of the ion exchange membrane at high pressure and temperature (a new membrane under development could minimise this problem).

Another concern which appears most critical for the mixed RFCS is the necessary quality of water produced by a static alkaline fuel cell and used by an acid electrolyser. The use of a cartridge deioniser bed seems mandatory at the present time which means additional maintenance or refurbishment. Furthermore, a water/gas separation is mandatory in the case of the PEM electrolyser with a similar volume of the two phases (not yet developed for working in micro-gravity).

However, European experience/background which pertains to recirculating KOH electrolysers has been abandoned for the RFCS application due to an unacceptable overall mass penalty, system complexity and bad estimated reliability. Only Dornier has begun (one year ago) preliminary laboratory research and development on static KOH electrolysers, whereas superior experience is available with PEM electrolysers used for submarine applications.

As a consequence, although technical preference is given to the static KOH RFCS, some additional activities with laboratory testing of hardware directed to a decision concerning the selection of the electrolyser should be necessary as a first step in the third study phase. Then the second part of this technology phase will be devoted to a detailed system engineering and optimisation of the selected configuration with, in addition, development work on the main components (electrolyser, pressure controller and storage units). A detailed 'breadboard' specification/requirement should complete this phase.

A three year phase of 'breadboarding' (January 1992 - end of 1994) will follow with the development, construction and tests of a complete RFCS at the laboratory level to demonstrate/confirm the feasibility of the RFCS.

Journal of Power Sources, 29 (1990) 223 - 237

SOLID OXIDE FUEL CELLS — THE NEXT STAGE

BRIAN RILEY

Power Systems, Combustion Engineering, Inc., 1000 Prospect Hill Road, P.O. Box 500, Windsor, CT 06095 - 0500 (U.S.A.)

Fuel cells are electrochemical systems that convert the chemical energy of the reactants directly into electrical energy. Over the last quarter of a century a number of fuel cell concepts have been developed up to and including commercial size devices. These are categorized according to the type of the electrolyte used in the cell (Fig. 1). All of the devices burn fuel at the anode or negative electrode, and consume an oxidant at the cathode or positive electrode.

The five main varieties of fuel cells, listed in increasing order of operating temperatures are:

(a) Polymer electrolyte fuel cell (SPFC) approximately 80 °C.
(b) Alkali fuel cell (AFC) approximately 100 °C.
(c) Phosphoric acid fuel cell (PAFC) approximately 200 °C.
(d) Molten carbonate fuel cell (MCFC) approximately 650 °C.
(e) Solid oxide fuel cell (SOFC) approximately 1000 °C.

The technical status of one of these systems, namely the solid oxide fuel cell and its derivatives, is presented here with respect to the basic design concepts, the materials of construction and their fabrication processes.

The three main SOFC variations are:

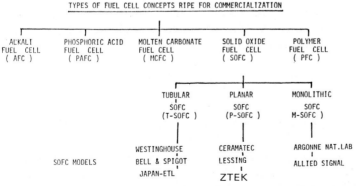

Fig. 1. An arbitrary selection of fuel cell concepts ripe for commercialization and of the various types of solid oxide fuel cell designs being vigorously investigated at this time.

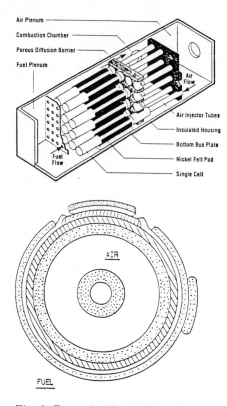

Fig. 2. Example of a solid oxide fuel cell tubular concept based upon the Westinghouse design.

(a) Tubular design (Fig. 2).
(b) Planar design (Fig. 3).
(c) Monolithic design (Fig. 4).

The above three SOFC designs differ only in cell geometry construction. The 'cell' is the repetitive electrochemical building block, connected in series and parallel, which form the 'stack' or unit of fabrication. The basic SOFC 'cell' consists of the following common parts:

(a) The anode.
(b) The electrolyte.
(c) The cathode.
(d) The interconnect or bipolar plate.
(e) The support tube (in the tubular design only).

A brief outline of the materials of construction of the above five components will highlight their commonality, for they are all based upon the same materials selection with minor dopant variations depending upon the fuel cell type and mode of fabrication. All are essentially ceramic materials,

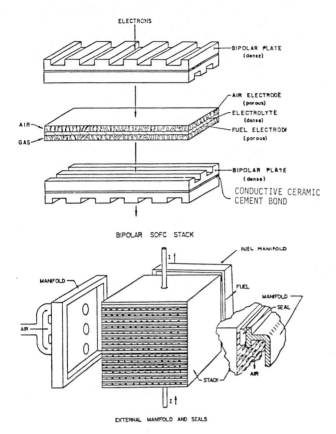

Fig. 3. Example of a solid oxide fuel cell planar concept based upon the Ceramatec and Lessing design.

synthesized and formed by conventional ceramic processes (Fig. 5). In each of the three SOFC types: the *anode* or fuel electrode is a porous cermet of nickel and an inert phase such as zirconia or yttria-stabilized zirconia; the anode is an electronic conductor with a projected current density in the range of 1 A cm^{-2}; it is fabricated usually as a mixture of nickel oxide and stabilized zirconia which is converted to the conductive cermet *in situ* within the cell. The fabrication process, however, will differ with the respective designs:

(a) Tubular design: The anode is slurry dipped onto the electrolyte, dried and sintered, or flame or plasma sprayed directly onto the already sintered electrolyte.

(b) Planar design: The anode is flame or plasma sprayed onto the fired electrolyte or bipolar plate, or in the pseudo-hybrid design, the anode will be tape cast or calender rolled in the green state, followed by assembly prior to debinding and sintering.

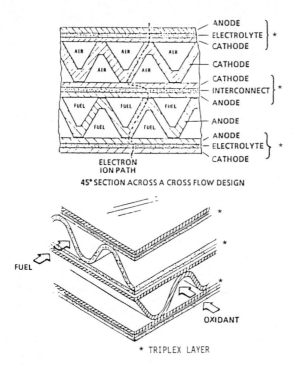

ANODE ⎫
ELECTROLYTE ⎬ *
CATHODE ⎭

CATHODE

CATHODE
INTERCONNECT ⎬ *
ANODE

ANODE

ANODE
ELECTROLYTE ⎬ *
CATHODE

ELECTRON
ION PATH

45° SECTION ACROSS A CROSS FLOW DESIGN

FUEL

OXIDANT

* TRIPLEX LAYER

Fig. 4. Example of a monolithic solid oxide fuel cell concept based upon the Argonne National Laboratory design.

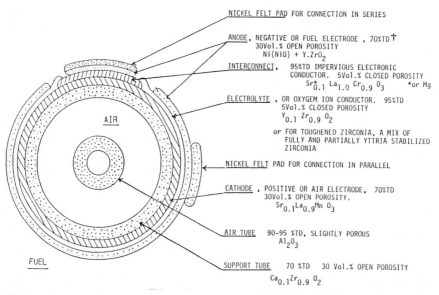

NICKEL FELT PAD FOR CONNECTION IN SERIES

ANODE, NEGATIVE OR FUEL ELECTRODE , 70%TD †
30Vol.% OPEN POROSITY
$Ni(NiO) + Y.ZrO_2$

INTERCONNECT, 95%TD IMPERVIOUS ELECTRONIC
CONDUCTOR. 5Vol.% CLOSED POROSITY
$Sr^*_{0.1} La_{1.0} Cr_{0.9} O_3$ *or Mg

ELECTROLYTE , OR OXYGEM ION CONDUCTOR. 95%TD
5Vol.% CLOSED POROSITY
$Y_{0.1} Zr_{0.9} O_2$

or FOR TOUGHENED ZIRCONIA, A MIX OF
FULLY AND PARTIALLY YTTRIA STABILIZED
ZIRCONIA

NICKEL FELT PAD FOR CONNECTION IN PARALLEL

CATHODE , POSITIVE OR AIR ELECTRODE, 70%TD
30Vol.% OPEN POROSITY.
$Sr_{0.1} La_{0.9} Mn O_3$

AIR TUBE   90-95 %TD, SLIGHTLY POROUS
Al_2O_3

SUPPORT TUBE   70 %TD   30 Vol.% OPEN POROSITY
$Ca_{0.1} Zr_{0.9} O_2$

AIR

FUEL

Fig. 5. List of solid oxide fuel cell components and their compositions using the Westinghouse concept as an example. Tube and layer thicknesses not to scale.   † %TD = percent of theoretical density.

(c) Monolithic design: the anode is tape cast or calender rolled, and assembled into the monolith in the green state prior to sintering.

The anode thickness is usually 100 - 200 μm with a theoretical density of 70%, the porosity being of the open type. The precise formulation of the anode is invariably a propriety formula, but will fall between 10 to 30% of the volume of the inert phase. The nickel is added as the nickel oxide and is reduced to the metal as the cell is heated in the reducing fuel atmosphere.

The ionic conducting *electrolyte* in most of the SOFC designs has a typical composition of $Y_{0.1}ZrO_{0.9}O_2$ or as is sometimes written: $(Y_2O_3)_{0.1}$-$(ZrO_2)_{0.9}$. Recent advances to toughen the zirconia, have introduced a two phase yttria-stabilized zirconia.

The trivalent Y^{3+} ion within the Zr^{4+} lattice produces anion vacancies and an oxygen ion conductor. The mobile ionic species within the electrolyte in the SOFC system are the O^{2-} ions as compared to protons (H^+) in the PAFC and SPFC and CO_3^{2-} ions in the MCFC (Fig. 6). The O^{2-} ion in the electrolyte travels from the cathode (air electrode) to the anode (fuel electrode) as does the carbonate ion in the MCFC. In contrast, the H^+ ion in the PAFC travels in the reverse direction from the anode to the cathode. All three current carrying species arrive at the appropriate electrolyte–electrode interface, or three phase boundary, and react with the gas phase within the porous electrode. The microstructural characteristic of this interfacial three-phase boundary is a critical parameter controlling the gas dynamics and the electrochemical efficiency of the fuel cell.

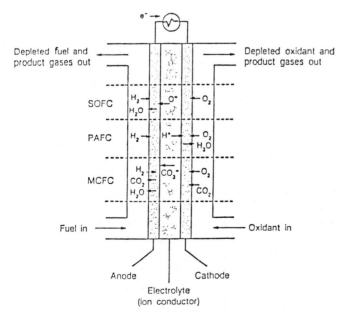

Fig. 6. Schematic illustration of various fuel cell ionic conductive species for the three main commercial designs.

A balance of properties and *in situ* performance witin the two porous electrodes and the impervious electrolyte must be maintained if the cell is to function economically (lifetime up to 40 000 h) and at the appropriate temperature (1000 °C).

The method of fabrication of the electrolyte will differ depending on the design concept, namely:

(a) Tubular design: Electrochemical vapour deposition from the mixed chloride gas by steam reduction (see Fig. 2), or plasma spraying mixed oxide powders onto the cathode (see Fig. 7).

(b) Planar design: Electrochemical vapour deposition of the mixed chloride gases, or plasma spraying the mixed oxides onto the porous electrode, or for hybrid design, tape casting or calender rolling the green tape.

(c) Monolithic design: Tape casting or calender rollling tapes to form anode–electrolyte–cathode triplex layers.

The impervious electrolyte tape is usually in the order of 50 - 75 μm in thickness with a theoretical density of 95%. The remaining 5% of the volume is the closed type. Recent advances, however, in SOFC materials selection have included 'toughened' zirconia ceramics based upon a two phase fully and partially stabilized yttria–zirconia system.

The *cathode* or air electrode for all three types is based upon the lanthanum manganite perovskite structure doped with strontium, of the general formula $Sr_xLa_{1-x}MnO_3$ where $x = 0.1$ to 0.2. The cathode is a p-type electronic semiconductor and, similarly to the anode, is a 70% theoretical density porous structure that must permit rapid diffusion of the air or oxygen to the electrolyte three phase boundary and subsequently to flush out the inert nitrogen. The thickness of the cathode will again be dependent upon the design and fabrication process, namely:

(a) Tubular design: The air electrode is slurry-dipped as a 1 mm layer onto the support tube, dried and sintered. In the advanced design for a self supporting cathode the thickness may be increased above 1 mm and be fabricated by mandrel extrusion, high pressure slip casting into moulds and/or centrifugal casting.

(b) Planar design: The air electrode thickness will be limited to the control and ability to retain the open porosity by a plasma or flame spraying process, *i.e.* 500 - 1000 μm. Hybrid design: tape cast or calender roll within a triplex layer. Co-extrude followed by sintering, could be an alternative to the spraying and will enable a better control of the porosity by the addition of organic pore formers.

(c) Monolithic design: Tape casting or calender rolling the cathode followed by co-rolling the triplex layers prior to corrugation, stacking and sintering. The thickness of the cathode is of the order of 40 to 50 μm.

The $SrLaMnO_3$ cathode tends to sinter at a lower temperature than the three sister layers in the monolith. The above mentioned methods for the fabrication of the cathode, *i.e.* slurry dipping, tape casting, calender rolling, slurry spraying and extrusion have included an organic pore former to control and retain the open porosity within the tape while in fabrication and in operation.

The *interconnect* or *bipolar plate* composition is based upon the electronically conducting lanthanum chromite perovskite structure doped with either strontium or magnesium. The following general formula applies to most designs $Sr_xLa_1Cr_{1-x}O_3$ where $x = 0.1$ to 0.2. In a similar fashion to the electrolyte, the interconnect is impervious to the fuel and oxidant gas and experiences (during operation at $1000\,°C$) both an oxidizing and a reducing environment. It should be appreciated how few candidates are available for the interconnect under these conditions where stability of both the crystal phase and the stoichiometry are essential. The interconnect thickness and mode of fabrication are again related to the SOFC design, namely:

(a) Tubular design: By electrochemical vapour deposition of the mixed chloride gases of Sr or Mg, La and Cr by steam reduction at a temperature between 1300 - $1600\,°C$ onto selected masked areas of the electrolyte (Fig. 5). The thickness is about $60\ \mu m$.

(b) Planar design (the bipolar plate is synonymous with the interconnect): The bipolar plate in the planar design is the support member of the SOFC cell housing the gas channels and acting as an impervious barrier. The process of fabrication is either by thick tape casting, calender rolling, injection molding or by classical wet or dry pressing. The 3 to 4 mm thickness will open up numerous fabrication techniques which are not applicable in thin film processing. The green or unfired plaque is either hard-fired followed by coating with the other electrodes or for the hybrid design bisque-fired then coated. The closing of the porosity in the bipolar plate will occur within the final shrinkage stage of co-sintering.

(c) Monolithic design: Tape casting or calender rolling. It has proved difficult to incorporate the interconnect into the 'one process' MSOFC procedure mainly due to the higher required sintering temperature to achieve the desired impervious state. It requires at least $1600\,°C$ to sinter the interconnect to 95% theoretical density and at this temperature the anode and cathode overdensify, closing off almost all their open porosity.

The *support tube* appears only in the Westinghouse and in the early bell and spigot design on which the Japanese ETL hybrid is based (Fig. 7). The support tube is the structural member of the tubular-SOFC system and also acts as the oxidant gas conduit (Fig. 2). The early composition was calcia-fully stabilized zirconia of the general formula: $Ca_xZr_{1-x}O_2$ where $x = 0.1$ to 0.2. The wall thickness of the T-SOFC support tube is about 1.5 to 2 mm and is highly porous to the oxidant gas. The initial porosity may be as high as 40% of the volume which within the further fabrication stages and in operation may close down to 25 - 30% of the volume. The support tube for the ETL bell and spigot design is made of porous alumina, which acts as the mandrel onto which the other SOFC components are plasma or flame sprayed using suitable masking devices.

With the Westinghouse design the support tube enters the T-SOFC fabrication line as a closed-ended bisque tube onto which the cathode, interconnect, electrolyte and the anode are sequentially deposited. The support tube experiences cumulatively five or six temperature (1200 to $1600\,°C$)

230

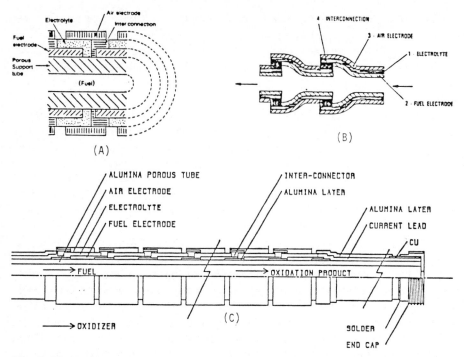

(A)

(B)

(C)

Fig. 7. Three examples of the solid oxide fuel cell 'bell & spigot' tubular concept based upon (A) Wade *et al.* [1]; (B) early Westinghouse design [2]; (C) Japan-ETL design.

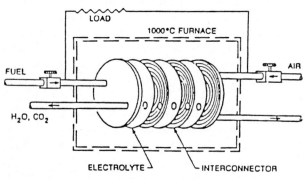

Fig. 8. Example of a solid oxide fuel cell planar model based upon the Ztek [3] design sponsored by EPRI in a 10 cell/1000 h test.

cycles within its fabrication life. Unfortunately the support tube in the original T-SOFC design accounted for 70% of the weight of the cell, which contributed to the lower energy density of the design relative to the monolith. The replacement of the calcia-stabilized support tube by a self supporting cathode will improve the energy density.

Although there are other SOFC designs (Fig. 8) the materials of construction are essentially those outlined above. Each of the various SOFC design concepts, now vigorously being developed in many countries for ultimate commercialization, exhibit their own characteristic problems in fabrication. The tubular concept, well advanced in field testing, uses expensive and very sophisticated electrochemical vapour deposition, masking and demasking, dip or spray coating to sequentially build up the various layers necessary for the electrochemical cell. The planar device relies on achieving controlled porosity in the plasma process and a high level of flatness for the building units to ensure edge seal integrity, minimum internal leakage crossover and most importantly to minimize interfacial resistance. The monolithic design will require all the skills and ingenuity of the ceramicist to match each of the mating layers through the stages of debinding and sintering, cooldown from the final thermal process, and throughout the thermal-ratchetting of cell operation. The debinding and sintering shrinkage of the four components and their thermal expansion coefficients must not differ by more than 5%. The converse of this will lead to delamination, cracking and reduction in cell performance.

The most adverse changes in SOFC performance are due to an increase in the internal resistance (IR), or IR drop across each cell. Due to the relatively low ionic and electronic conductivity of the oxide ceramic materials of choice, to reduce the IR drop, SOFCs are essentially thin film devices. The fabrication of thin layers in many thousand square meters in area has now become a specialized technology. Some of the specific processes of fabrication being:

(a) Slurry spraying.
(b) Slurry dipping.
(c) Tape casting.
(d) Calender rolling.
(e) Thin lamellar extrusion.
(f) Flame spraying.
(g) Plasma spraying.
(h) Electro and chemical vapour deposition.
(i) Plasma assisted CVD.
(j) Laser assisted CVC.

The first five process methods prepare a thin layer in the green state which will require further processing prior to the final sintering stage. The remaining five process techniques produce a high density layer, not necessarily requiring further treatment. However many of the sophisticated techniques used in the semiconductor device industry are being used in the development of the SOFC. Examples are screen printing, continuous sheet casting onto a moving substrate, masked CVD and laser assisted CVD. Likewise it is envisaged that over the next decade, parallel researches into superconductivity will contribute to the search for new or improved conductive SOFC materials.

In order for a technology to be evaluated against a background of a long term energy programme, based partially on fuel cells, it is prudent to know the economics of the alternative concepts, to thoroughly understand

and quantify the materials and processes of production, and to evaluate the ultimate need and size of the SOFC module required by the customer. These requirements will reflect the variability of performance under the customer's field conditions, and for the manufacturer, the degree of sophistication and availability of the materials of construction. For the manufacturer an under estimation of the materials availability or reproducible quantities of a chosen process can have profound repercussions to the price of electricity and the cost per kilowatt installed.

The 'degree of reproducibility' of a fabrication process precis the technical and developmental requirements necessary for the SOFC manufacturer to address over the next five years. Unfortunately, unlike fossil and nuclear power plants, fuel cell plants based on solid oxide technology are built up from a large number of small fuel cell stacks. It is impossible to make a large fuel cell from green ceramic components — the manufacturing building unit. Therefore can one million of a component, identical in all respects, be made from knowledge and experience of making only a thousand?

The manufacturing building unit is the fabricated component which travels through the processing line from raw materials to the final product and which is capable of being quality assured as an acceptable entity. For each of the three SOFC concepts reviewed, the manufacturing unit is:

(a) Tubular design: The single power generating tube of about 200 - 300 W capacity.

(b) Planar design: The single plate cell prior to stacking of 300 - 500 W capacity.

(c) Monolithic design: The stack, built from 50 - 200 cells in the green state. The capacity of the stack in the sintered state will be 25 - 50 kW.

The manufactured unit fabricated for shipping, *i.e.* the assembled SOFC with bus bars, conduits, manifolding, insulation etc., envisaged for the three fuel cell types is:

(a) Tubular design 10 - 15 kW module.

(b) Planar design 20 - 25 kW module.

(c) Monolithic design 0.5 - 1.0 MW module.

These units will be assembled into larger commercial size plants on site.

For the cell fabricator, it is important to know the relationship between the size of the shipping unit and the components which are capable of being quality assured prior to release. This will then reflect the cost of the unit. Present numbers reflect an arbitrary target of $1500/kW SOFC stack installed which is related to a SOFC fabrication cost of $350 - $450 for the fabrication of the ceramic fuel cell stack prior to connecting to the manifolding and gas conduits. Both these cost figures will restrict the raw materials of manufacture to about $50 to $70 per kW. The above numbers are target costs for the 90s and should be used as 'drivers' for the fuel cell technologists to create a mind-set in scale-up procedures from the laboratory scale to the industrial commercial plant.

To achieve a hypothetical fuel cell production of six-sigma, the statistician's term to define virtually errror free performance, both process and product must be rigorously quality controlled/quality assured. The axiom — "the product is a natural extension of a process, and in many cases they are inseparable" — will be amply applicable to the ceramic fuel cell fabrication. Extending this axiom further, not only is the product an extension, but both the process and product are likewise an extension of the raw materials and their process and source of supply. The manufacturer of the future may be faced with a multitude of series and parallel material flow system networks each of which must be quality controlled/quality assured prior to the acceptance of a component.

Advanced materials have drawn researchers around the world to evaluate new and sophisticated ways of making materials that ordinarily would not be economical or capable of large-scale production. For example, the conventional method of making the mixed perovskite for conductive electrodes was by mechanically mixing the oxides or carbonates of the cations. To break up the bisque after reaction calcining, percussion ball milling was invariably required. Although the early electrode materials were made by this procedure, the method is not applicable to fuel cell electrode fabrication, unless stringently quality controlled. The solid state method invariably leads to poor sinterability, hard strongly bonded agglomerates, inhomogeneity, mixed oxide phases, abnormal and bimodal grain size, poor reproducibility and shrinkage control, imprecise cation stoichiometric ratios and unstable mechanical and electrical properties within the cell. Regrettably this solid state method seems to be the only process available to date with batch and lot yields in the thousands of kilograms per day capacity.

At this time small 1 to 10 kg batches of the mixed cation electrode and interconnect materials have been successfully prepared using the Pechini technique based on the co-precipitation from the mixed cation polymeric precursor. The Pechini method uses citric acid and cation salts with ethylene glycol. This produces a resin like solid which after charring at about 400 - 800 °C, results in a fine mixed perovskite powder. Unfortunately this process does not lend itself to 1000 kg/day batches due to the high cost of ethylene glycol and to an excessive off gasing exothermic stage. However derivatives of this excellent synthesis technique may be capable of scale-up. SOFC ceramic components will possibly require the following parametric definitions to accurately and precisely define the requirements of a specification of the 'starting' materials:

(a) The starting powder stoichiometry which, within the fabrication process and under the cell operating conditions, will achieve the desired product stoichiometry.

(b) The starting powder morphology. The standard sub-set of this being (i) particle size and distribution; (ii) surface area.

(c) The chemistry of the starting powder.

In previous years the above three parameters were enough to 'index' the starting powders and for the industrial ceramicist to tailor the variables of the process to fit the properties of the raw materials. Today however these are not enough. SEM, EDEX, powder shape, emission spectroscopy, X-ray diffraction of the powders and possibly others are needed to 'fingerprint' the raw materials or to detect the degree of differences and similarities between prior and subsequent lots. Most raw powder specifications are relative rather than absolute, being compared and related to powder lots in the experimental trials which 'work'. The theoretical specification for the pure powder is not a practical proposition in the commercial field. However the hypothetical requirements of the SOFC ceramicist to achieve the six sigma process will inevitably lead scientists to search for the 'perfect powder'.

For the SOFC requirements this possibly consists of a nanophase sub-micron, mono-distribution of particles whose mean deviation from the norm is no greater than −5% relative. The 'perfect powder' compositional stoichiometry likewise should not differ from the required mean by, at least, the cumulative limits of error of all the evaluations determined on the individual constituents.

In the search for flaw-free ceramic layers in the SOFC fabrication, the properties of the starting materials are of paramount importance in the control of the layer integrity through the debinding and sintering stages. In parallel, similar careful research must be directed to process control throughout all the stages of fabrication, especially in the sometimes forgotten debinding of the green state.

Thin layer fabricators are now using multi-component organic systems in the ceramic preparations to achieve the desired match with other layers. As many as six to eight organics may be used in the slip or 'leather' stage. These organics will all be required to crack or volatilize between room temperature and 500 °C without leaving any carbonaceous residue to alter the stoichiometry by a carbothermic reaction. The gas phase evolution in such a mixture is many hundreds of volume percent. The fuel cell ceramic infra-structure must be capable of accommodating this gas evolution without distortion or delamination. The following will be the tools of the quality assurance analyst in the monitoring of the fabrication process.

(a) The recording differential dilatometer: To measure the shrinkage throughout the debinding and sintering stages.

(b) The differential scanning calorimeter: To assess any phase changes in the constituents which could cause delamination on thermal cycling.

(c) Thermogravimetric analysis: To measure the weight loss in the debinding and sintering stages.

(d) Thermovapouremetric analysis: To assess the gaseous evolution species on burnout of the organics.

(e) The X-ray diffractometer: To assess the crystallographic phases present in the starting materials and in the final product.

(f) X-ray CAT* scan or MRI**: To assess the layer or structural integrity of the assembled stack.

---

*Computerized Axial Tomography. ** Magnetic Resonance Imaging.

However it must be realized that the sintering process of a ceramic 'starts' at one degree above room temperature. Above this all the individual constituents begin the intricate process of decomposition, rearrangement and finally the mechanism of compaction. Of the many new procedures open to the ceramicist to supplement these, the following are worthy of interest:

(a) The supply of ultra fine powders in the Ångstrom range (nanophase materials).

(b) The use of microwave processing of ceramics.

In previous years the researcher had thought little of the sources of the materials of his research if and when his endeavors reach the commercial production scale. The premise that a materials market source will always develop in sympathy with the demand may not necessarily be true in the 90s. As an example, for a SOFC 200 MW/year manufacturing capacity the following will be required:

(a) Tubular design: 4 million SOFC generator tubes.

(b) Planar design: 2 million SOFC generator plates.

(c) Monolithic design: 750 000 SOFC plates.

The active surface area for a 200 MW plant will be in the order of 50 - 80 000 m^2 and this will represent in bulk raw materials weight for the tubular design:

(a) $CaZrO_2$ ~ 500 000 kg.

(b) Y and Zr chlorides   ~ 50 000 kg.

(c) NiO   ~ 50 000 kg.

(d) $SrLaMnO_3$ ~ 200 000 kg.

(e) $SrLaCrO_3$ ~ 100 000 kg.

For the monolithic design the weights of the raw materials will be somewhat lower, namely:

(a) $Y.ZrO_2$ ~ 170 000 kg.

(b) $SrLaMnO_3$ ~ 100 000 kg.

(c) $SrLaCrO_3$ ~ 50 000 kg.

(d) NiO   ~ 50 000 kg.

(e) Organics   ~ 100 000 kg.

These material requirements are three orders of magnitude greater than the present day capacity 'prepared' on the 1 - 10 kg scale, to fit the SOFC fabricators specifications. The consistancy and reproducibility from lot to lot, and batch to batch is well below the standards required for large-scale production. This is forcing the developmental technologist to change the process to fit the raw materials. This is a dangerous situation of a 'moving' target-moving platform which for the production quality assurance manager is a veritable nightmare. The next stage in the development of the SOFC technology will possibly be in the direction of requiring the SOFC fabricator to perform critical path analysis on each of the components, their fabrication process and their raw materials. This is to detect, prior to the decision point of commercialization, weak areas in the system, the process, or stages within the process which when viewed against a backcloth of present day technology are incapable of scale-up.

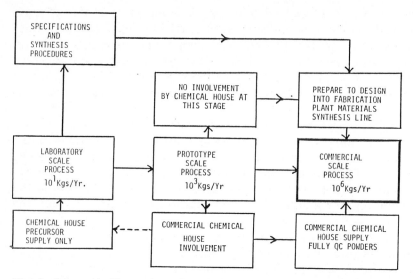

Fig. 9. Schematic illustration of the development of a solid oxide fuel cell commercial source supply of raw materials based upon a quality control dominated fabrication process.

Possible solutions to the variable source starting materials will lie in achieving a good working relationship with a commercial chemical house at the earliest possible time in the experimental programme (Fig. 9). In the development of a preparative or fabrication process, large temperature/ time/pressure coefficients of reaction should be avoided, *i.e.* do not base a technology on a process with a control of +2 at 500 °C for 15+ min. These values will invariably create either a very expensive product or a variable supply. It is essential to develop a fabrication procedure with built in fail-safe quality assurance staging points which will prevent a "bad" component progressing up a fabrication line undetected. The cumulative worth of a SOFC component as it progresses through the process increases exponentially. The worth of a monolithic SOFC 50 kW stack changes from less that $1000 worth of raw materials to about $20 000 after sintering. One single layer within the stack can fail the entire stack.

A detailed evaluation of the manufacturing flow charts should include a lookout for a 'material virus' — a component, part, source or design which when 'injected' into a system can cause profound damage at a later and invariably a critical point. The virus can be subtle in that it will not affect the component in which it is in, but will fail or cause deterioration in another. An example of a potential material virus would be the poor control of the manganese stoichiometry in the air electrode mix. Through the fabrication and operation of a fuel cell, the cathode may perform well, yet the excess manganese could migrate to the electrolyte and cause an internal electronic short.

For the future SOFC ceramicist the following are areas of possible interest within the preparation and synthesis of SOFC materials, which over the next decade are worth paying attention to for their influence on scale-up:

(a) Mixed oxide powders by new polymeric precursors.

(b) Mixed cation sol–gel synthesis.

(c) EVD and CVD plasma and laser assisted processing.

(d) Microwave processing of the precursor and raw materials.

(e) Microwave processing throughout the entire ceramic stages.

(f) Ultra-fine colloidal particle research and the colloidal mill nanophase material.

(g) High intensity ultrasonics in preparative processes.

(h) High intensity and high speed water jets.

(i) Fluidized bed-microwave processing of microspheres.

(j) Explosive or supercritical drying or precipitation processes.

(k) Emulsion ion-exchange co-precipitation of mixed cations.

(l) Controlled precipitation from organic and mixed solvents.

(m) Compositional and stoichiometric inhomogeneity control.

(n) Advances in tape casting and calender rolling of thin films.

Of these, the microwave treatment of ceramics will probably have the most profound and significant impact on the entire ceramic research and development technology and the ceramic industry in general from the refinement of the raw materials to the end product. This is an area to be carefully watched over the next decade.

The ceramicist-developer-fabricator has now, and into the 90s, an impressive number of analytical and preparative tools of the trade which can be applied to monitor the entire materials processing and the analysis of the process. Many of the tools are 'real time' quality control systems which provide instant feedback to the plant operator of the conditions and in some devices potential problems within the process. The next stage in the SOFC for the fabricator will be to assess the process analytics available to the process and to apply these in a prudent fashion and frequency to achieve that six sigma state within the SOFC development.

SOFC have a tremendous future in their ability to generate clean and efficient power and their capability of adaption for large-scale power generation by modular design and by microcogeneration using down sized modules. At this time the SOFC electrochemistry is reasonably well understood; the materials of choice are in the process of final screening for the next jump; the next stage in the material science of the SOFC will be to increase the volume of these materials to the prototype and then to the commercial level.

For the SOFC corporate organizations however, fuel cell technology may follow in the same path of two sister technologies, namely those of semiconductors and superconductors, with respect to the hi-tech maxim: 'They who control the materials control the technology'.

References

1 C. J. Wade, E. F. Sverdrup and A. D. Glasser, in G. Sandstede, *Electrocatalysis to Fuel Cells*, University of Washington Press, Seattle, WA, 1972.
2 Fuel cells, *DOE/METC 86/0241*, Morgantown, WV, 1986.
3 Ztek Co., Lincoln, MA, U.S.A.

Journal of Power Sources, 29 (1990) 239 - 250

THE RENAISSANCE OF THE SOLID POLYMER FUEL CELL

KEITH PRATER

*Ballard Power Systems, Inc., 980 West 1st Street, Unit 107, North Vancouver, BC,
V7P 3N4 (Canada)*

Historical background

The solid polymer fuel cell (SPFC) or SPE (a trademark owned by Hamilton Standard) fuel cell was first developed by General Electric (GE) for NASA in the 1960s [1]. It consists of two porous electrodes, which are lightly catalyzed on one surface, bonded on either side of a thin sheet of a hydrogen ion-conducting polymer, the solid polymer electrolyte. The backs of the porous electrodes are contacted by plates which contain channels through which a fuel gas is supplied to the back of the anode and an oxidant gas is supplied to the back of the cathode. Electrical contact to the electrodes may be made through these fluid flow field plates.

The perceived advantages of the SPFC for space applications were its high energy density compared to batteries, the absence of corrosive, liquid electrolytes, the relative simplicity of the stack design, and the ruggedness of the system.

The technology initially suffered from a limited operating lifetime, due to degradation of the membrane electrolyte. By 1964, GE had developed membranes based upon the cross-linking of styrene-divinylbenzene into an inert fluorocarbon matrix. SPFCs based upon these membranes exhibited lifetimes of about 500 h and were satisfactory for their use in seven Gemini missions.

In the mid-1960s, GE, working with DuPont, adapted DuPont's Nafion for use in the SPFC. This fully fluorinated material exhibited a substantially improved operating lifetime — in excess of 57 000 hours [2]. GE used Nafion in 1968 for the Biosatellite mission. At this point, long operating lifetime and low maintenance requirements could be added to the advantages of the SPFC.

By then, NASA had selected the alkaline fuel cell for use in the Apollo program. There had been a perception within NASA that the polymer electrolyte was intrinsically resistive and that the requirement for a higher power density fuel cell system for Apollo could be better met by the alkaline fuel cell. This, for all practical purposes, put the SPFC on the shelf for space applications for the next 20 years.

GE chose not to pursue commercial applications of the SPFC, probably because of the perceptions that, as compared with the phosphoric acid fuel cell, the SPFC was more expensive (expensive membrane and high platinum loading) and more sensitive to CO poisoning. The latter concern was seen as precluding the use of common carbon-containing fuels with the SPFC and thus severely limiting its market potential. With the exception of limited work under the sponsorship of Los Alamos National Laboratory, solid polymer fuel cell technology lay dormant until about 1984.

The beginning of the renaissance

In 1983, the Canadian Department of National Defence (DND), in association with the National Research Council, determined that solid polymer fuel cell technology might satisfy some of the growing military power needs and have commercial applications as well, if it could be re-engineered for terrestrial applications and at a lower cost. In early 1984, Ballard began a two-year contract with DND to acquire SPFC technology and to evaluate its potential.

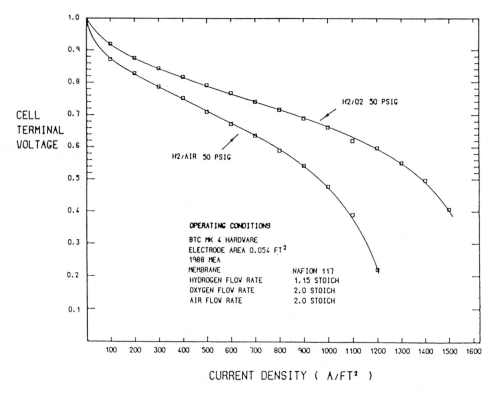

Fig. 1. Comparison of single cell performance on H_2/O_2 to performance on H_2/air.

Air as the oxidant

The initial focus of the Ballard/DND program was the development of stack hardware which would operate effectively on air, as well as on pure oxygen. This required improvements in the distribution of air to the back of the porous cathode, the removal of product water, and the manifolding of cells in a multi-cell stack.

Figure 1 shows the present performance of single cells operating on hydrogen/oxygen and on hydrogen/air at the same pressure. This cell had an active electrode area of 0.054 ft^2 (46.5 cm^2) and used Nafion as the membrane electrolyte. Note that, at a given terminal voltage, the current (or power) produced using air is about 70% that obtained using pure oxygen at the same pressure. In this experiment, the air flow rate was five times that of the oxygen flow rate, so that the total oxygen passing through the cell was the same in each case.

Performance obtained in single cells is often difficult to maintain in multi-cell stacks. Figure 2, however, demonstrates comparable performance in a 54-cell stack. Again, performance on air is about 70% that obtained on pure oxygen.

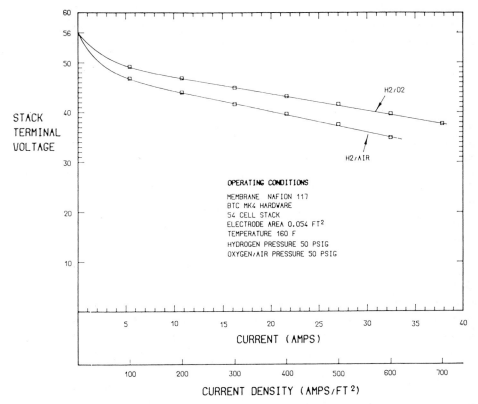

Fig. 2. Comparison of 54-cell stack performance on H$_2$/O$_2$ to performance on H$_2$/air.

Carbon-containing fuels

The second aspect of the Ballard/DND program was the demonstration of acceptable performance on reformed hydrocarbon fuels. These gas mixtures typically contain 70 - 80% hydrogen, 20 - 30% carbon dioxide and 0.1 - 1.0% carbon monoxide. The SPFC is essentially insensitive to the CO_2 in the gas stream, but very sensitive to the CO. Ballard developed [3] a selective oxidation process which is used to pre-treat the reformed fuel gas stream before it enters the fuel cell. Table 1 shows the performance of a 0.05 ft^2 electrode area cell when operated on pure hydrogen, hydrogen/CO_2, hydrogen/CO_2/CO, and hydrogen/CO_2/CO treated to selectively oxidize the CO. The anode contained a CO-tolerant catalyst, as well as platinum. The data shown reflect performance after 24 h of operation.

Note that, while the addition of 25% CO_2 to the fuel gas had a relatively minor effect on performance, the further addition of 0.3% CO dramatically reduced performance, even with the CO-tolerant catalyst. In fact, it was not possible to pass the reference 400 A/ft^2 at any meaningful voltage with CO present in the fuel gas. When fuel gas containing CO was passed through the selective oxidation process before entering the fuel cell, fuel cell performance was essentially identical to that observed with only CO_2 present.

Table 2 shows that this performance was retained in a 12-cell stack on the same fuel combinations. These data show performance after 24 h. Even after 500 h, performance on treated reformate was around 90% of that for pure hydrogen.

TABLE 1

Synthetic reformate/air performance MK 4 single cell/30 psig/185 F/Nafion 117

Fuel gas	Voltage (V)	Current density (A/ft^2)	Power (% H$_2$/air)
H$_2$	0.71	400	100
H$_2$/25% CO$_2$	0.68	400	96
H$_2$/25% CO$_2$/0.3% CO	0.71	200	50
Treated fuel	0.67	400	95

TABLE 2

Synthetic reformate/air performance MK 4 12 cell stack/30 psig/Nafion 117/400 A/ft^2

Fuel gas	Voltage (V)	Power (% H$_2$/air)
H$_2$	8.22	100
H$_2$/25% CO$_2$	8.05	98
Treated fuel	7.94	97

Cost reduction

Materials replacement/reduction

Cost reduction was addressed from two points of view — reduction in materials cost and improvement in performance. The Ballard cell uses low cost graphite for the fluid flow field plates, instead of the niobium used in the NASA cell plates. Ballard also determined that baseline performance could be obtained with significantly lower platinum loading on the electrodes than the 8 mg/cm^2 per cell which has been the standard. Los Alamos has recently reported good performance down to about 0.4 mg/cm^2 per cell [4].

Performance improvement/the membrane electrolyte

The most significant reduction in cost has resulted from improvements in performance. In 1987, Ballard received a new ion-conducting polymer membrane from Dow Chemical. The Dow membrane is a sulfonated fluorocarbon polymer [5], like Nafion. As shown in Fig. 3, when placed in the Ballard cell hardware, the Dow membrane produced four times the current (and power) at the same operating voltage as that obtained when using a Nafion membrane electrolyte [6]. The polarization data in Fig. 3 were obtained after 120 h of continuous, stable performance at 4000 A/ft^2 (4.3 A/cm^2).

As shown in Fig. 4, this performance was retained in a six-cell stack. The polarization data in Fig. 4 were obtained after 20 h of operation at

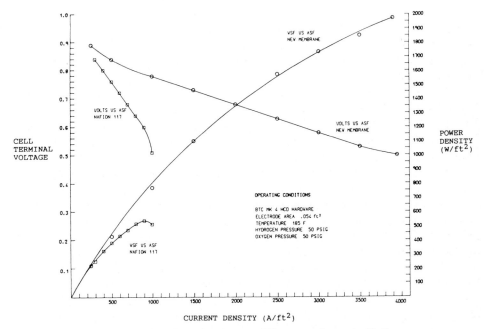

Fig. 3. Comparison of single cell performance of Dow membrane to Nafion.

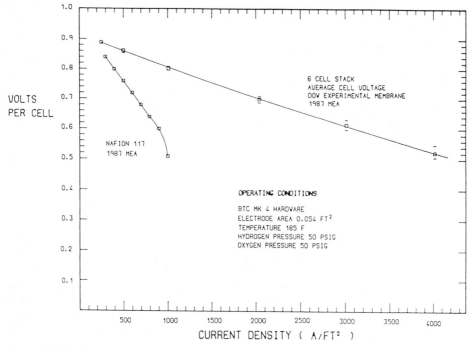

Fig. 4. Multi-cell stack performance on Dow membrane.

4000 A/ft². These data indicate that both the cell-to-cell gas manifolding and the thermal management design of the cell are sufficient for continuous operation at these current and power densities.

By increasing the operating pressure to 100 psig (7.8 atm) and the temperature to 215 °F (102 °C), current densities in excess of 6000 A/ft² have been obtained as shown in Fig. 5.

Performance improvements have also been achieved with Nafion 117. Figure 6 demonstrates a roughly 50% improvement in limiting current density in a single cell using the same Nafion 117 as the electrolyte. This improvement results from changes in the fabrication procedure for the membrane/electrode assembly.

Dow has provided several versions of its membrane for evaluation. These samples have varied in thickness, in equivalent weight, and, presumably, in other fabrication variables [6]. Figure 7 shows the variation in performance which has been obtained with the various samples as compared with the recent Nafion performance. The thicknesses of these membranes are tabulated in Table 3.

Based upon the data available to Ballard, it is not possible to determine which variables are most important to performance. It is clear, however, that there is substantial flexibility in the fabrication process which should allow for performance optimization for a variety of applications. It is also clear that the Dow material is, in all samples tested, superior to Nafion for this

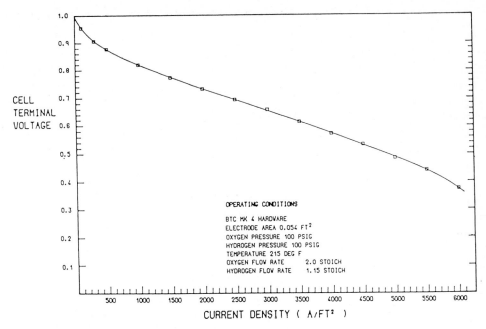

Fig. 5. High temperature, high pressure performance on Dow membrane.

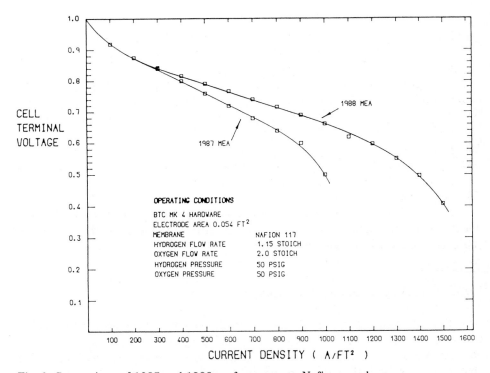

Fig. 6. Comparison of 1987 and 1988 performance on Nafion membrane.

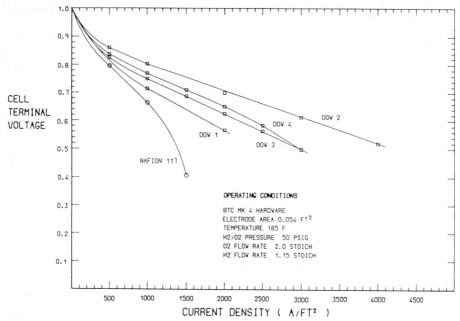

Fig. 7. Comparison of the performance of various Dow membranes with that of Nafion 117.

TABLE 3

Membrane thickness

Membrane	Thickness (in)
Dow 1	0.0067 - 0.0071
Dow 2	0.0035 - 0.0039
Dow 3	0.0052 - 0.0055
Dow 4	0.0063 - 0.0066
Nafion 117	0.0083 - 0.0087

application. The improvement in performance derived from the Dow membrane represents a substantial reduction in the size, weight and cost for an SPFC delivering a given amount of power.

Performance improvement/stack scale-up

Further size, weight and cost reductions were obtained by increasing the electrode size and, in the process, reducing the amount of peripheral material required in the stack. The original MK 4 hardware had an active electrode area of 0.054 ft^2 (7.8 in^2, 50.2 cm^2) but required graphite plates

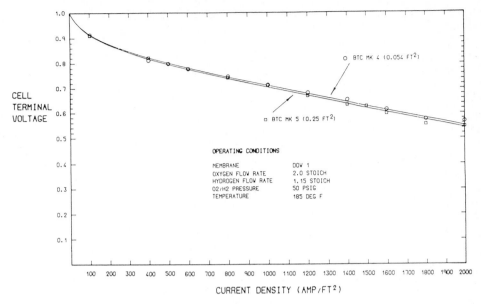

Fig. 8. Comparison of MK 4 and MK 5 hardware performance on H_2/O_2 with the Dow membrane.

and membrane electrode assemblies which were 5 in × 5 in (12.7 cm × 12.7 cm). The MK 5 hardware has an electrode area of 0.25 ft^2 with graphite plates and membrane electrode assemblies only 8 in × 8 in (20.3 cm × 20.3 cm). The hardware scale-up has increased stack power by a factor of 4.63, while increasing stack cross-sectional area and stack volume by only a factor of 2.56. As seen in Fig. 8, the scale-up was completely linear for hydrogen/ oxygen using the Dow membrane. Perhaps more impressive, the scale-up was also linear for hydrogen/air performance as shown in Fig. 9.

Commercialization

In addition to advancing the state of the technology, Ballard is committed to commercializing solid polymer fuel cells. To that end, in 1987 Ballard delivered a 2 kW hydrogen/oxygen fuel cell consisting of two-54 cell MK 4 stacks containing Nafion electrolytes to Perry Energy Systems in Florida. The unit was housed in a container 1 ft in diameter and 2 ft long and was intended to power an unmanned submersible. Shortly thereafter, Ballard delivered an identical unit to the U.K. Royal Navy for evaluation.

The Perry unit has since been retrofitted with a single MK 5 Nafion-based stack in place of the two MK 4 stacks. The upgraded unit can provide up to 4.5 kW in the same volume as the original unit. This MK 5 unit provides the entire power requirements for a two-man submersible, which is now undergoing sea trials.

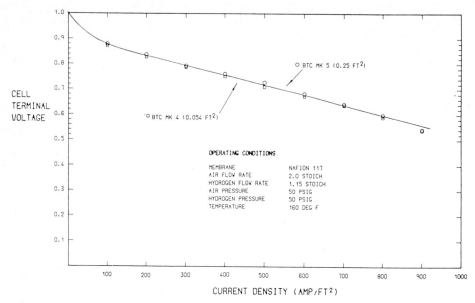

Fig. 9. Comparison of MK 4 and MK 5 hardware performance on H_2/air with Nafion 117 membrane.

Fig. 10. MK 5 20-cell stack.

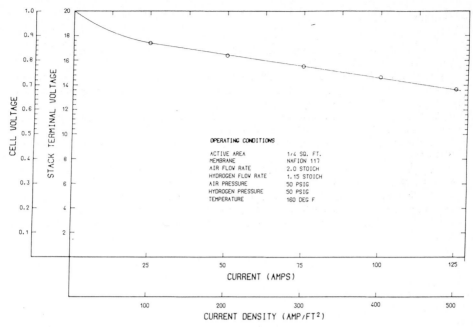

Fig. 11. Performance of 20-cell Nafion-based MK 5 stack on H_2/air.

Daimler-Benz is currently evaluating a 20-cell MK 5 hydrogen/air stack, which produces 2 kW using Nafion electrolytes. That stack, which is approximately 1 ft × 1 ft × 1 ft (30.5 cm × 30.5 cm × 30.5 cm), is shown in Fig. 10. The performance of that stack on hydrogen/air is shown in Fig. 11.

A 10 kW hydrogen/air MK 5 system using the Dow membrane will be delivered in September, 1989 for installation in the Dow chlor-alkali plant in Sarnia, Canada. This represent the first of a number of 10 to 50 kW installations planned for Europe, Japan, the U.S. and Canada over the next 18 months.

Conclusions

The substantial improvement in SPFC performance which has been demonstrated over the past two years has resulted in a reappraisal of the possible applications of this technology and the potential markets which are now open to it. Major companies, such as General Motors and Siemens, have significant programs in this area. It now appears that, after 20 years on the shelf, the solid polymer fuel cell is ready for commercialization and may open substantial markets for fuel cells which are not possible with any other fuel cell technology.

References

1 A. J. Appleby and F. R. Foulkes, *Fuel Cell Handbook*, Van Nostrand Reinhold, New York, 1989, and refs. therein.

2 *Solid Polymer Electrolyte Fuel Cell Technology Program, Test Report BU#1 and BU#2 (1.1 ft²)*, Contract NAS 9-15286, Direct Energy Conversion Programs, General Electric Co., TRP-76, May 1980.

3 D. Watkins, K. Dircks, D. Epp and A. Harkness, Development of a small polymer fuel cell, *32nd Int. Power Sources Symp., June 1986*.

4 E. A. Ticianelli, C. R. Derouin and S. Srinivasan, *J. Electroanal. Chem.*, *251* (1988) 275 - 295.

5 B. R. Ezzell, Low equivalent weight sulfonic fluoropolymers, *Eur. Patent Applic. 88 106 319.2*, (April 20, 1988).

6 G. A. Eisman, The application of a new perfluorsulfonic acid ionomer in proton-exchange membrane fuel cells: new ultra-high current density capabilities, *Ext. Abstr. Fuel Cell Technology and Applications, Int. Seminar, The Netherlands, Oct. 1987*, p. 287ff.

Journal of Power Sources, 29 (1990) 251 - 264

FUEL CELLS FOR TRANSPORTATION

ROSS A. LEMONS

Los Alamos National Laboratory, Los Alamos, NM 87544 (U.S.A.)

The 150th anniversary of the Grove fuel cell is an outstanding opportunity to review the status of fuel cell technology and its potential applications. One of the most exciting and most challenging of these applications is in transportation. As shown in Fig. 1, a fuel cell powered vehicle is conceptually simple. The fuel cell provides an efficient means of converting chemical energy to electricity. If a fuel cell with adequate power capacity could be housed within the engine compartment of a vehicle, its electrical output could be used to drive an electric motor for propulsion as well as all of the electrical ancillary equipment of a modern vehicle.

In the early 1980s, Los Alamos National Laboratory developed a set of computer models and performed assessments of the performance of potential fuel cell powered vehicles [1, 2]. The conclusion of these studies was that with reasonable improvements in power density a fuel cell powered vehicle could deliver comparable performance to an internal combustion engine (ICE) powered vehicle at approximately twice the energy efficiency (*i.e.* km/l on the same fuel) and with negligible pollution. The fuell cell system could also provide comparable range and refueling time to the ICE. These have been the main deficiencies of battery powered vehicles.

The motivations to look for an alternative technology to ICEs are growing rapidly. These include: the need to improve the efficiency of fossil fuel utilization both to extend the duration of these valuable resources and to reduce the rate of carbon dioxide production; the need to reduce air and noise pollution, particularly in urban areas; and the need in many countries to reduce the dependence of transportation on petroleum.

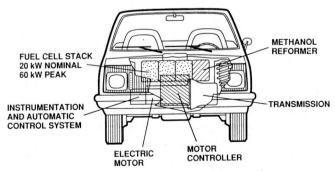

Fig. 1. Cut away view of a conceptual layout for a passenger car fuel cell propulsion system.

0378-7753/90/$3.50

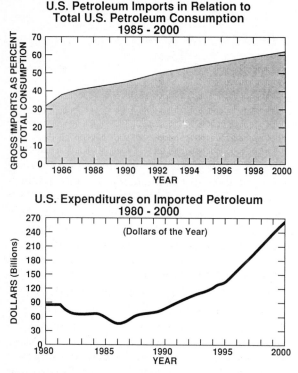

Fig. 2. The top graph shows the projected growth of United States petroleum imports in relation to total U.S. petroleum consumption. (Sources: U.S. DOE for historical and DRI Energy for projected.) The bottom graph shows the projected U.S. expenditures on imported petroleum. (Sources: U.S. Department of Commerce for historical and forecasts derived from DRI Energy projections.)

To illustrate the problem of petroleum dependence Fig. 2 shows the projected petroleum imports to the United States and their projected cost as a function of time. By the year 2000 the United States will be importing more than 60% of its petroleum at a cost of over $260 billion per year. In the United States, most of this petroleum is used for transportation. As shown in Fig. 3, the transportation sector in the United States already requires more petroleum than is domestically produced. The bulk of this (also shown in Fig. 3) is being used in automobiles. Both the efficiency of fuel cells and their ability to use other sources of fuel could help to alleviate this petroleum dependence problem.

Pollution is also a growing problem. As shown in Fig. 4, transportation is a major contributor to NO_x pollution and many of the other components of urban smog. Fuel cells, particularly those operating at low temperature, produce negligible amounts of NO_x and much lower levels of other pollutants and particulates than internal combustion engines. They are also intrinsically quiet.

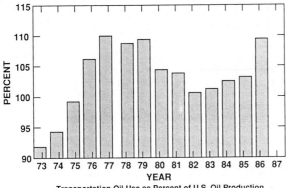

Transportation Oil Use as Percent of U.S. Oil Production

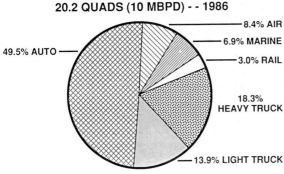

Petroleum Use By Transportation Mode

Fig. 3. The histogram at the top shows that transportation is a growing fraction of U.S. petroleum consumption and that this fraction has exceeded domestic production. (Source U.S. DOE.) The pie chart at the bottom shows that automobiles consume nearly 50% of the petroleum consumed by the U.S. transportation sector. (Source U.S. DOE.)

The greenhouse effect is potentially one of the most severe long term environmental problems we face. As other fossil fuels, such as coal, are substituted for petroleum this problem will worsen. Transportation is a significant contributor to CO_2 production, as shown in Fig. 4. The fuel cell with its higher efficiency can reduce the rate of CO_2 production in comparison with ICEs burning the same fuel.

These are among the motivations for exploring the potential of fuel cells for transportation applications. To realize this potential, major challenges must be met. The requirements for transportation applications are extremely stringent. The fuel cell must have sufficient power density to meet the performance specifications of the vehicle and to fit within the available space. It must be sufficiently inexpensive to compete with internal combustion engines on an economic basis. It must be able to start rapidly and to respond quickly to changes in power demand. It must be safe and reliable. When using hydrocarbon fuel, the fuel cell must tolerate the carbon dioxide

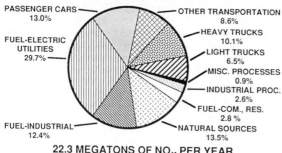

1986 NO$_x$ EMISSIONS IN THE U.S.

22.3 MEGATONS OF NO$_x$ PER YEAR

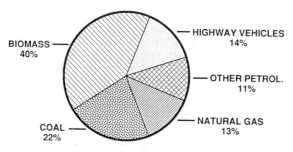

1985 CO$_2$ EMISSIONS IN THE U.S.

CARBON DIOXIDE 8 x 10^9 METRIC TONS

Fig. 4. The top chart shows the sources of nitrogen oxide emission in the U.S. More than 1/3 is produced by transportation. (Source: U.S. Environmental Protection Agency.) The bottom chart shows the sources of man made carbon dioxide emission in the U.S. In 1985 highway vehicles produced approximately 14%. (Source: U.S. Environmental Protection Agency.)

that is produced in the reforming process or it must be able to use the hydrocarbon directly.

All of the major types of fuel cells have been evaluated for their ability to meet these requirements. Some of the important features are summarized in Table 1. As might be anticipated, no one type is superior to the others in all respects. Indeed, one type may be more suitable than another for a specific application. For example, in ships or locomotives in which frequent on/off cycling is not required, the high-temperature fuel cells, such as solid oxide or molten carbonate, present advantages. They allow more flexibility in fuel selection and may be used without a reformer. The high-grade waste heat is more easily used in a thermally integrated system. Conversely, the low-temperature fuel cells, such as polymer electrolyte or alkaline, may be a better choice for passenger cars to which rapid start-up and wide power range are important.

TABLE 1

Fuel cell comparison

Fuel cell type	Operating temperature (°C)	CO_2 tolerant	Power density H_2−air (W/cm^2)	CO tolerance	Reformer required
Solid oxide	1000	yes	0.20 - 0.27	good	no
Molten carbonate	600	yes	0.19 - 0.24	good	no
Phosphoric acid	150 - 205	yes	0.20 - 0.29	fair	yes
Alkaline	65 - 220	no	0.15 - (4.3)	poor	no
Solid polymer	25 - 120	yes	0.2 - 0.9	poor	yes

Recently, major advances have been achieved in fuel cell technology, which make the application to transportation much more viable. These advances include large increases in power density, major reductions in intrinsic cost and improved system design.

For example, Argonne National Laboratory in the United States has been developing a monolithic solid oxide fuel cell that has the potential for very high power density [3]. These cells operate at ~1000 °C, potentially eliminating the need for a reformer. The monolithic design employs the same materials used in the tubular solid oxide design. These materials provide excellent thermal-expansion matching, which is critical for cycling to the operating temperature of the cell. For the monolithic cell, the materials are deposited by tape-casting techniques. Because the layers are thin, high spcific power is potentially available. Current work is focused on improving the fabrication procedures, improving the materials integrity, and on reducing interface resistances in stacks.

Much interest in alkaline fuel cells for transportation has been evident in Europe, where hydrogen is more widely available. Although it cannot withstand CO_2 in the fuel stream, the alkaline fuel cell offers very high power density and efficiency. In Belgium, Elenco has developed a 15 kW alkaline fuel cell for powering an electric test van [4]. This company is particularly interested in commercially operated vehicles (city buses, refuse trucks, vans, etc.) for transporting people or goods in urban areas.

At Los Alamos National Laboratory with support from the United States Department of Energy, we have focused our attention on the polymer electrolyte fuel cell. This technology offers a combination of characteristics, which we feel make it the leading contender for passenger vehicles. It provides high power densities with values over 2 W/cm^2 on H_2 and O_2 as reported by Ballard Technologies Corporation in Canada [5]. It can tolerate CO_2 in the fuel stream thus allowing the use of reformed hydrocarbon fuel. It can be self-starting at temperatures above ~20 °C. Low-cost structural materials can be used because of the low operating temperature and reduced corrosion. And the solid character of the electrolyte facilitates sealing and safety of the fuel cell stack.

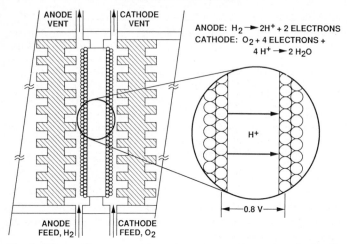

DEVICE PRODUCES ELECTRICAL ENERGY FROM HYDROGEN OXIDATION

ANODE: $H_2 \longrightarrow 2H^+ + 2$ ELECTRONS

CATHODE: $O_2 + 4$ ELECTRONS + $4 H^+ \longrightarrow 2 H_2O$

Fig. 5. Schematic cross section of a polymer electrolyte fuel cell.

The limitations of the technology have been high platinum content, expensive membranes, poor CO tolerance, water management problems and difficulty in thermally integrating with a reformer. The objective of the Los Alamos program has been to solve these problems.

The cross section of a single polymer electrolyte fuel cell is shown schematically in Fig. 5. Typically the cell consists of graphite bipolar plates which are pressed against the membrane–electrode assembly. These plates have a manifold of grooves which distribute the reactant gases to the electrodes. They are also sufficiently electrically conductive to pass the generated current to the adjacent cell.

The membrane–electrode assembly is the electrochemical heart of the system. On the anode or hydrogen side, hydrogen gas is catalytically dissociated according to the reaction

$$H_2 \longrightarrow 2H^+ + 2e^-$$

The hydrogen ions pass through the polymer electrolyte to the cathode or oxygen side of the cell. There they are combined catalytically with oxygen and electrons from the adjacent cell to form water, according to the reaction

$$4H^+ + O_2 + 4e^- \longrightarrow 2H_2O$$

The polymer electrolyte, from which the cell derives its name, is a remarkable ionically-conducting, plastic-like material in the form of a membrane 50 to 175 μm thick. Membranes sold commercially by DuPont under the Nafion® trade name and several types of experimental membrane manufactured by DuPont, Dow and other companies are used. These materials are perfluorosulfonic acids (teflon-like fluorocarbon polymers with side chains ending in sulfonic acid groups). To maintain their protonic conductivity the membrane must contain sufficient water. In operation, maintaining the correct water content is one of the key aspects of polymer electrolyte fuel cell design.

In order for the electrochemical reactions to take place at useful efficiency they must be catalyzed. To date, platinum has proven to be the best catalyst for both the hydrogen oxidation (anode) and the oxygen reduction (cathode) reactions. To function, the catalyst must have access to the gas and must be in contact with both the electrical and protonic conductors.

This may be achieved by pressing the platinum catalyst into the membrane and then contacting it with a porous electrode that provides both electrons and gas access. The deficiency of this approach is that relatively large amounts of platinum must be used to obtain adequate catalytic activity. Typically, 4 mg of platinum are used on each square centimeter of electrode area. The problem is that platinum is much too expensive to be used in this quantity for a consumer application. With such loading, an automobile would require up to $10 000 worth of platinum.

To solve this problem, at Los Alamos the use of supported platinum catalysts similar to those used in liquid electrolyte fuel cells is being investigated. The supported catalyst consists of 2 to 5 nm diameter Pt particles on the surface of fine carbon particles. This greatly increases the effective surface area of the platinum. In much of this work electrodes manufactured by Prototech Inc. in which these carbon particles are mixed with teflon and bonded to a porous carbon cloth have been used. Unfortunately when such electrodes are pressed against the polymer electrolyte membrane only those few catalyst particles at the surface make effective contact with the protonic conductor.

This problem has been solved by impregnating the supported catalyst electrode with protonic conducting material. This is achieved by covering the surface with a solution of solubilized Nafion®. As shown schematically in Fig. 6, if the solution is the appropriate concentration, it will leave a

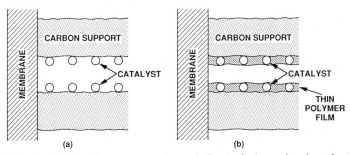

Fig. 6. Schematic representation of the technique developed at Los Alamos for use of supported platinum catalysts in polymer electrolyte fuel cells. In (a) the catalyst particles are ineffective because protons from the membrane cannot migrate across the carbon surface. By impregnating the pores of the electrode with a thin layer of protonic conductor (b) the supported catalyst becomes highly effective.

protonically conducting film on the catalyst particles that connects them to the membrane without blocking gas access. This technique has proven to be extremely effective. Performance comparable to the high catalyst loading cells has been achieved with approximately 1/20 the platinum, reducing the catalyst cost for an automobile sized fuel cell to a few hundred dollars.

The performance characteristic of a fuel cell is shown schematically in Fig. 7. As the current drawn from the fuel cell increases the voltage falls. Ideally at low currents the voltage would be near the reversible potential, 1.23 V. In practice, the catalyst does not function perfectly and produces activation losses of 0.2 to 0.3 V. At higher currents, ohmic losses due to the finite resistance of the cell contribute an additional 0.1 to 0.3 V of loss. At sufficiently high current densities (1 to 4 A/cm^2) reactant gases can no longer be supplied to the catalyst sites at sufficient rates to sustain the reaction and the voltage drops rapidly to zero. In this regime the power output from the cell begins to fall. Each of these voltage losses also corresponds to a loss of efficiency, since the efficiency is simply the ratio of the cell voltage to 1.48 V.

For comparison, Fig. 7 shows the current–voltage characteristic of a polymer electrolyte fuel cell using low catalyst loading electrodes. This cell

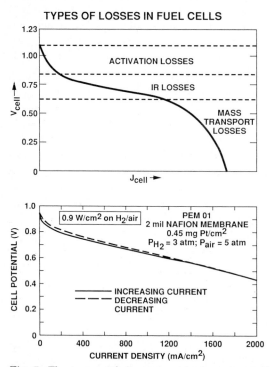

Fig. 7. The top graph is a conceptual representation of the current–voltage characteristic of a fuel cell, showing the major contributors to voltage loss. The bottom graph shows an experimentally measured current–voltage characteristic of a polymer electrolyte fuel cell using low platinum loading electrodes.

TABLE 2

Fuel cell parameters for a passenger car

Net continuous power	20 kW		
Net peak power	60 kW		
Operating points			
Peak	0.50 V	1.8 A/cm^2	0.9 W/cm^2
Continuous	0.75 V	0.4 A/cm^2	0.3 W/cm^2
Stack size			
Active area	500 cm^2		Diameter 25 cm
Cross section	1000 cm^2		Diameter 35 cm
Cell thickness	0.5 cm		
No. cells	133		
Stack length	66 cm		Total 75 cm
Stack voltage (nom.)	100 V		
Stack volume	0.072 m^3		
Stack density	1.0 g/cm^3		
Stack weight	75 kg		Total 100 kg

TABLE 3

Comparison of hydrogen storage technologies

Technique		cal/g	cal/cm^3
Gas	steel cylinder (200 atm)	510	240
Liquid	cryogenic H$_2$	4250	2373
Hydrocarbon	n-octane	11400	8020
	methanol	5340	4226

is capable of delivering a power density of 0.9 W/cm^2 at a current density of 2 A/cm^2 using hydrogen and air. A practical power plant is made by stacking a large number of such cells in series. Table 2 shows a set of design parameters for a stack of 133 such cells with an active area of 500 cm^2 each. The full stack would provide 100 V at 200 A at the continuous operating point of 20 kW. The power output could be increased to 60 kW by increasing the current density to 1.8 A/cm^2. With individual cell thicknesses of 0.5 cm the stack would be approximately 75 cm long and 35 cm in diameter with a weight less than 100 kg including assembly hardware and pressure housing. Such a stack could easily meet the power density requirements for a compact car giving performance comparable to that provided by a standard 4 cylinder ICE power plant.

    The choice of fuel is one of the key factors in the design of a fuel cell propulsion system. Most fuel cells operate best on pure hydrogen and oxygen. Usually air can be used directly as the oxidant with little loss of performance. Pure hydrogen can be carried aboard a vehicle either as compressed gas or as a cryogenic liquid. However, as shown in Table 3, the volumetric energy density of pressurized hydrogen is poor. The energy density of liquid hydrogen, though better, is still less than 1/3 that of gasoline, and the technology for storing small volumes of liquid hydrogen is not well developed. Hydrocarbons by comparison provide a simple way to store hydrogen at high density. In some types of fuel cell, hydrocarbons can be used directly. In others they must be chemically converted to hydrogen by a catalytic process called reforming. As a reformed fuel, methanol is particularly attractive. It can be efficiently converted to H_2 and CO_2 by reacting it with water on a Cu–Zn catalyst at ~180 °C.

In the United States particularly, methanol is a likely fuel to replace gasoline in the transportation sector as petroleum reserves are depleted. Methanol can be synthesized from a variety of domestic resources including coal, natural gas and biomass. It is a liquid fuel that is easily integrated into the existing distribution system, and has about half the energy density of gasoline. Since a fuel cell propulsion system has about twice the energy efficiency of an internal combustion engine, it can achieve comparable range on a tank of methanol to that of an ICE burning gasoline from a tank of the same size.

To use methanol in a polymer electrolyte fuel cell propulsion system, one must have an on-board reformer that can meet the vehicle requirements. These include rapid start-up and quick transient response to meet the changing fuel demands. Conventionally reformers have not been designed to meet these requirements, often taking an hour or more to start up and many minutes to change output significantly.

At Los Alamos reformer designs which may solve these problems have been investigated [6]. One approach is shown schematically in Fig. 8. In

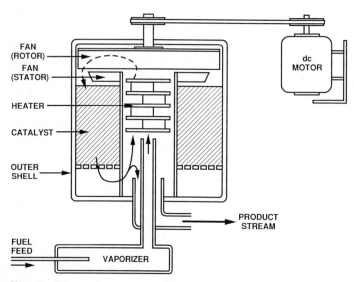

Fig. 8. Schematic cross section of a recirculating methanol–water reformer developed at Los Alamos.

this design an internal fan is used to recirculate the reformate through the catalyst bed. In this way, the catalyst bed can be kept uniformly at its optimal temperature, and heat can be rapidly injected to increase the reforming rate.

One of the problems with methanol–water reformers is that the reaction is often not complete, leaving small concentrations (approximately 1%) of carbon monoxide in the fuel stream. This carbon monoxide is an extremely effective poison to the fuel cell catalyst. As shown in Fig. 9, even a few parts per million CO produce a substantial degradation in the fuel cell performance, particularly at high current densities.

One approach to solve this problem is to reduce the carbon monoxide content of the reformate. This has been achieved by selectively oxidizing the CO to CO_2 [5, 6]. To achieve this, the reformate passes through a small reactor containing a platinum catalyst. As shown in Fig. 10, by injecting a small amount of oxygen or air into this reactor, the CO content can be dramatically reduced with relatively little hydrogen consumption.

At Los Alamos, recent work has shown that the performance degradation caused by trace amounts of CO still remaining in the reformate can be restored by injecting small amounts of air at the anode [7]. For example, Fig. 11 shows that the severe performance degradation caused by 100 ppm CO can be completely restored by injecting 2% air into the anode feed.

Much has been accomplished over the last few years in solving some of the key problems that impeded the use of fuel cells in transportation. The polymer electrolyte fuel cells in particular are looking increasingly attractive for this application. The power density has been greatly increased, the platinum requirements have been dramatically reduced, and solutions to carbon monoxide poisoning have been found. Much work still remains to be done before this technology can become commercial. However, the viability

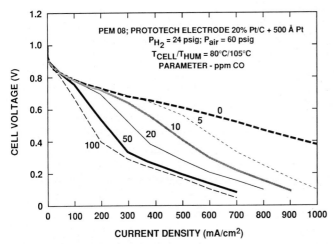

Fig. 9. This graph shows the performance losses caused by trace amounts of carbon monoxide in the fuel stream.

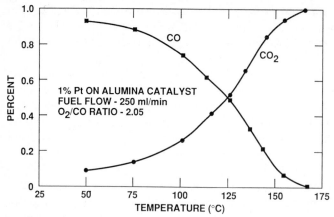

Fig. 10. This graph shows that the residual carbon monoxide in the methanol reformate can be effectively converted to carbon dioxide by post-reformer selective oxidation at a temperature of ~160 °C.

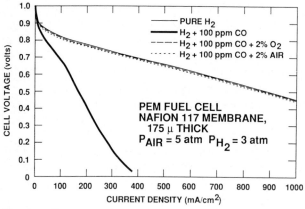

Fig. 11. This graph shows that residual catalyst poisoning by trace amounts of CO can be completely eliminated by injecting ~2% air into the fuel cell anode.

of the technology is sufficiently well established to begin to solve the problems of system integration.

From the system perspective, the fuel cell stack is just one of a large number of subsystems that must be integrated to produce an effective propulsion system. The complexity of this task is indicated in Fig. 12.

These difficult problems are beginning to be addressed. The U.S. Department of Energy (DOE) and the U.S. Department of Transportation (DOT) have initiated a cooperative multiyear program with industry for the research, development and demonstration of a fuel cell/battery hybrid bus system for urban passenger transport [8]. In its early development, this system will be based upon a phosphoric acid fuel cell (PAFC), sized to provide the average power requirements of the bus, and a battery pack

FUEL CELL PROPULSION SYSTEM

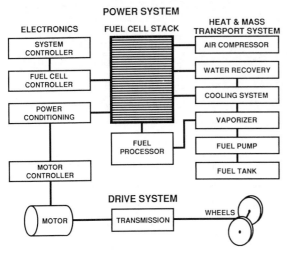

Fig. 12. Block diagram of a fuel cell propulsion system showing the large number of subsystems that must be integrated.

capable of meeting the peaking requirements [9]. The PAFC system was selected for the early phases of this project because it is currently the most highly developed fuel cell system. It is expected that the polymer electrolyte fuel cell will replace the PAFC in the latter stages of the program.

In conclusion, fuel cells offer an attractive alternative power source for transportation applications. They provide over twice the efficiency of internal combustion engines and can operate on non-petroleum-based fuel, such as methanol. Significant advances have been made in increasing the power density and reducing the intrinsic cost of fuel cells, but much work remains before they can be successfully integrated into vehicles on a large scale.

References

1 D. K. Lynn, J. B. McCormick. R. E. Bobbett, S. Srinivasan and J. R. Huff, Design considerations for vehicular fuel cell power plants, *Proc. 16th Intersoc. Energy Conversion Engineering Conf., Atlanta, GA, Aug. 9 - 14, 1981*, Vol. 1, The American Society of Mechanical Engineers, New York, 1981, pp. 722 - 725.

2 J. R. Huff (ed.), Fuel cells for transportation applications, *Los Alamos National Laboratory Rep LA-9387-PR*, June 1982.

3 D. C. Fee *et al.*, Monolithic fuel cell development, *Abstr. 1986 Fuel Cell Seminar*, Oct. 26 - 29, 1986, Courtesy Assoc. Inc., Washington, DC, pp. 40 - 47.

4 H. Van den Broeck, G. Van Bogaert, G. Vennekens, L. Vermeeren, F. Vlasselaer, J. Lichtenberg, W. Schlösser and A. Blanchart, Status of Elenco's alkaline fuel cell technology, *Proc. 22nd Annual Meeting IECEC, 1987*, American Institute of Aeronautics and Astronautics, Inc., pp. 1005 - 1009.

5 D. Watkins, D. Dircks, D. Epp and C. De la Franier, Canadian solid polymer fuel cell development, *Fuel Cell News, VI* (2) (June) (1989) S1 - S4.

6 N. E. Vanderborgh, Methanol fuel processing for low-temperature fuel cells, *Abstr. 1988 Fuel Cell Seminar, Oct. 23 - 26, 1988*, Courtesy Assoc. Inc., Washington, DC, pp. 52 - 56.

7 S. Gottesfeld and J. Pafford, *J. Electrochem. Soc., 135* (1988) 2651 - 2652.

8 *Program Plan for Research and Development and Demonstration of a Fuel Cell/ Battery Powered Bus System*, U.S. Department of Energy, Office of Transportation Systems, Jan. 7, 1987.

9 H. S. Murray, DOT fuel-cell-powered bus feasibility study, *Los Alamos National Laboratory Rep. LA-10933*, June 1987.

Closing Remarks

GROVE ANNIVERSARY FUEL CELL SYMPOSIUM — CLOSING REMARKS

A. J. APPLEBY

Center for Electrochemical Systems and Hydrogen Research, Tees/Texas A&M University, College Station, TX 77843 (U.S.A.)

The Grove Anniversary Fuel Cell Symposium was a successful and wide-ranging meeting. It was very well attended, with a wide selection of delegates from the United Kingdom, the other European Community Countries, European Organizations, the United States, Canada and Japan. The participants included members of academia, government and industry, the latter representing both the developers and the potential users of fuel cells. It was a particular privilege to address the distinguished audience in the historic auditorium of the Royal Institution, dating from the year 1800, from the very spot where Michael Faraday and other famous nineteenth century scientists delivered their many lectures.

To begin with some remarks concerning the device itself, it should be noted that a fuel cell power generator is much more than just the fuel cell proper, *i.e.*, the d.c. power unit containing the individual stacked cells consuming hydrogen and oxygen. This part is the 'fuel cell' as assumed by the electrochemists. The d.c. power produced by this fuel cell generator must be converted to a.c., some of which is required for auxiliary systems. Water must be condensed from the fuel cell exhaust, and any spent anode gas must be used effectively. The hydrogen must be produced from primary fuel, for example methane. Using the phosphoric acid system as an illustration, the pure water reaction product will be advantageously employed for cooling the cell stack, raising steam for reforming of natural gas fuel. The endothermic heat of reforming will be provided by anode tail gas. Reformer exit gases must then be water–gas shifted to remove CO, at the same time producing more hydrogen. The plant must finally be characterized by efficient heat recovery throughout.

Overall, the plant may occupy five times the footprint or the volume of the fuel cell d.c. generator stack itself, so that the cost of the latter may only be a small portion of the total cost. Indeed, it may be only about 20% in a mature unit. The cost of the other items, which are in most cases standard chemical engineering equipment, and state-of-the-art electronic equipment, may appear to be closer to practical mass production than the fuel cell stack itself. However, the entire system today still consists of a collection of components, or rather spare parts, including a chemical engineering subsystem, quasi-integrated with the fuel cell stack, which is in turn integrated with necessary electronics, which are presently not off-the-shelf, for control and for d.c.–a.c. conversion.

0378-7753/90/$3.50

Taking as an example a fuel cell system using natural gas fuel, the chemical engineering portion of the plant for the conversion of the feedstock to a hydrogen/carbon dioxide mixture has typically been a collection of parts taken from the ammonia industry. Even though this is usually considered to be state-of-the-art, it is so for a different application. A future fuel cell power generator cannot use the same ground rules as a traditional ammonia plant. It would operate under a different set of circumstances, with different economics. In a plant converting natural gas to a chemical product, such as ammonia, a different set of conditions must operate from those in a plant converting the energy available in natural gas to a different form of energy that must be economically competitive.

The very existence of an energy conversion plant will depend on its immediate economic competition. A power plant using a fuel cell is still a thermodynamic engine, with a cascade of Carnot heat-machines (pressure–temperature devices) combined with the primary electrochemical direct energy convertor. The theoretical efficiency of a thermal engine operating at a given heat sink temperature is the same as that of a fuel cell operating on the same fuel at the same temperature. This fact is intuitively implied by the second law: the maximum work that one can extract from a fuel is the free energy available in that fuel. Both devices reject a combustion product at a sink temperature, resulting in a practical work loss. The main reason why the thermal engine usually has a lower efficiency than a fuel cell is that it cannot operate at a heat source temperature that even approximates to that which can be theoretically produced by combusting the fuel. This results from materials limitations in the heat source heat exchange system. Secondly, any practical cycle has thermodynamic losses.

Similarly, the fuel cell cannot use all of the free energy available in the fuel, because of inevitable inefficiencies. These are analogous to the high temperature loss in a thermal engine. However, the high temperature thermal engine losses due to the irreversible $T\Delta S$ terms are normally greater than those in practical fuel cells. An idealized practical combination is a fuel cell combined cycle. In this, the fuel cell produces work at the thermal engine heat source temperature, so that fuel cell waste heat can be thus recovered. The high operating temperature of the fuel cell reduces irreversibly its losses, and addition of the thermal bottoming cycle compensates for the lower free energy available from the fuel at the high fuel cell operating temperature. The theoretical thermal efficiency (η) of the fuel cell alone is equal to

$$\eta = \Delta G_1/\Delta H \tag{1}$$

where ΔG_1 is the free energy available in the fuel at the operating temperature of the fuel cell (*i.e.*, at the heat source temperature of the thermal engine, T_1) and ΔH is the heat of combustion of the fuel. A generalized formula for the maximum work available in any energy conversion device, whether it be a fuel cell operating alone, an ideal thermal engine operating at the maximum theoretical temperature (the combustion temperature of the fuel), or a fuel cell thermal engine combination, is given by the expression

$$\eta = \Delta G_2 / \Delta H \tag{2}$$

where ΔG_2 is the free energy available in the fuel at the heat sink tempera-ture, T_2. For an isothermal fuel cell operating alone, T_1 and T_2 are of course identical. The corresponding expression for a so-called Carnot-limited thermal engine operating between source and sink temperatures T_1 and T_2 is given by

$$\eta = (\Delta G_2 - \Delta G_1)/(\Delta H - \Delta G_1) = (T_1 - T_2)/T_1 \tag{3}$$

At the spontaneous fuel combustion temperature, $\Delta G_1 = 0$, and eqn. (3) becomes identical to eqn. (2). If a fuel cell operating at T_1 is used as a topping cycle for the thermal engine, its theoretical efficiency is given by eqn. (1), and the fraction of waste heat available for further conversion at T_1 is $(1 - \Delta G_1/\Delta H)$. The overall theoretical efficiency of the combination is given by the sum of eqns. (1) and (3), the latter multiplied by the fraction of waste heat available. Rearrangement of this expression gives a result for the overall efficiency of the combination equal to eqn. (2). It should be pointed out that ΔG_2 in this expression is not the standard value, but that for the practical fuel conversion (utilization) desired.

Thus, if a fuel cell is used as a topping cycle in combination with a thermal engine, their theoretical losses cancel each other, and the combina-tion behaves ideally. The practical losses (*i.e.*, irreversibilities) in a high tem-perature fuel cell are low, and a thermal engine can be designed to operate at typical heat source temperatures corresponding to the operating temperature of the fuel cell. The fuel cell and the thermal engine are therefore comple-mentary devices, and a practical fuel cell 'black box'* would be such a combination. Even a low temperature fuel cell, for example phosphoric acid, benefits from a thermal bottoming cycle. A system recovering waste heat from the cells to operate a turbocompressor is an example.

The dominant theme stressed by successive authors in the Symposium revolves around the low environmental impact of fuel cells. Therein lies the political impact of the technology at the present time. Their high efficiency, though it is certainly important for the future, is less emphasized as a cost benefit in the commercialization phase than their extremely low level of tropospheric chemical and acoustic pollution. However, assuming that the fuel cells consume hydrogen, even if it is derived from fossil fuels, their high efficiency means lower carbon dioxide emissions, thus a lower greenhouse impact on global warming. These points were repeatedly stressed during the meeting. It should be noted that hydrogen used in a fuel cell produces no NO_x in the oxidation process, whereas hydrogen consumed in an internal combustion engine does. Thus, the later solution will not eliminate air pol-lution in cities: in fact, pollution will be almost as bad as that resulting from burning methanol in the internal combustion engine. Since hydrogen is never likely to be an inexpensive fuel, its use in the fuel cell rather than in the

---

*Or perhaps, due to its low environmental impact, a 'green box'.

internal combustion engine should be encouraged by the former's much greater fuel efficiency. Perhaps one can hope that future legislative credits will encourage the development of the hydrogen fuel cell automobile, because of its low social cost and because of the low capital cost for its infrastructure, which would result from its low energy requirements per kilometer.

One cannot fail to be impressed by the Japanese commitment towards fuel cell commercialization. Their industry has determined that the part of the device requiring new experience, *i.e.*, the fuel cell stack, will be manu-facturable using mass production methods. It therefore will be made at costs that are a small multiplier of those for the materials alone. For the phos-phoric acid system, the semi-finished materials costs for the cell stack should correspond to about $80/kW without catalyst, and $170/kW with catalyst. For other fuel cell technologies, *e.g.*, molten carbonate and solid oxide electrolyte systems, it is perhaps too early to tell. However, the molten carbonate system uses fundamentally simple materials technology, and should therefore not be costly when fully developed. Since the solid oxide system uses advanced ceramics with their associated difficulties in process-ing, it is less easy to make predictions about its future. The fundamentally simple alkaline and fluorinated acid polymer electrolyte cells, which can essentially use plastics as major construction materials, are perhaps even less defined from the viewpoint of final production cost, simply because they have been less studied from the viewpoint of the impact of the learning curve impact on their technology.

The Japanese developers and users are conducting numerous field tests of units of different technology and origin at the present time, and are thus building up a corpus of very valuable knowhow. This will give them the possibility of determining the factors influencing reliability, as well as cost, under real-world conditions. Such information, which unfortunately is not available elsewhere, either in Europe or in the United States, will place their industry in a privileged position for the manufacture of commercial units, particularly from the viewpoint of simplification and manufacturability consistent with reliability and ease of maintenance.

The continued commitment of the U.S. Agencies to fuel cell develop-ment is impressive. This has continued to be true whether they are public such as DOE, or semipublic bodies such as EPRI and GRI. Their interest in promoting the technology in the United States is still real, despite the disap-pointment resulting from the waning of interest by potential electric utility users. This has resulted from economic factors, particularly to the present low cost of oil-based fuel or natural gas, much less than that predicted in the early 1980s. In addition, the attitude of the electric utilities has not been aided by a perceived pricing policy on the part of developers to sell intro-ductory units at costs beyond their real economic value, even for niche market applications. This is particularly true when the reliability and main-tenance cost of early units is uncertain, and likely to be high. Thus, the cost of electricity of such units would be dominated by high capital and (probably) O&M costs. Their improved efficiency and low environmental impact compared with competitive technologies in similar unit sizes are less important when fuel costs are low, as they are at present. However, as stated above, their good neighborliness should swing the balance in the future, away from the major competition, which is represented by advanced gas turbines and combined cycles. For on-site integrated energy systems (OS-IES) producing electricity and waste heat, which can be used for space heating and absorption-cycle air conditioning, the economic picture is brighter, since this market becomes economic at higher capital costs than those for electric utility generation. In consequence, a substantial number of U.S.-developed PC-25 200 kW OS-IES units are on order from International Fuel Cells. Some of these are being installed in Japan, along with a wide range of units developed domestically in that country.

It is encouraging that the fuel cell has again been taken up in Europe, after the pioneering work there from the nineteenth century to the late 1960s. The Netherlands has now a major program on the development of the molten carbonate high-temperature system, originally developed in that country by Broers and Ketelaar from the early 1950s onwards until about 20 years ago. Some major advances have already been made in the Netherlands in the past twelve months, notably in the development of new anode materials (with seven patent applications) and in co- or counterflow sheet-metal bipolar plate design. In Italy, a similar program is now under way, under the leadership of ENEA, with Ansaldo as prime developer. This follows the referendum held on future nuclear power in Italy in the wake of Chernobyl. No further conventional reactors will be constructed, and only power derived from advanced fail-safe reactors, or fusion systems, when these are available, will be permitted in the future. In this regard it seems a pity that there is so little interest in the development of fuel cell systems in the United Kingdom, since so many of the pioneering efforts took place here, as this Symposium has amply demonstrated. Elsewhere in Europe, the most effective effort is that on new materials, led by the EEC. This topic will be discussed later.

As stated above, the major attraction of the fuel cell power plant will be its low environmental impact. This includes lower carbon dioxide emissions compared with its rivals, especially if it is cascaded with a heat recovery cycle to obtain units with thermal efficiencies in the order of 60%, or possibly higher. When used in vehicles, it will make a major impact on the pollution problem in cities. In this respect, the United Kingdom and many European countries are 15 years behind the United States and Japan in enacting pollution controls for gasoline powered vehicles. Even these are not enough, especially in communities such as the Los Angeles basin. The situation in Mexico City, Rome and Athens is similar. Volatile reactive organic compounds, carbon dioxide and NO_x can be legislated against, but there is a limit to what can be achieved with the internal combustion (IC) engine. For example, gasoline contains enough aromatic hydrocarbons, particularly benzene, to be a cancer hazard. The fuel cell power plant can, in contrast, eliminate these pollutants. For use in vehicles, at least based on foreseeable technology, the fuel cell power plant will require hydrogen fuel. This hydrogen might be carried on board (compressed, as liquid hydrogen, or as a hydride), or transported in the form of a hydrogen fuel carrier, particularly methanol or possibly ammonia. The use of the carrier fuel will require a load-following reactor inside the vehicle, which will be bulky and complex if high efficiency is to be achieved. An easier solution will be the use of a conversion system, probably methanol to hydrogen, in the gasoline service station or at the local level. Whether methanol will be used on a large scale as a clean-burning fuel will depend, in the United States at least, on pending clean-air legislation. In the long term, it will probably happen as oil supplies diminish and as fuel becomes based on clean coal technology. However, the oil companies are actively lobbying against methanol at present, ostensibly on safety grounds, but in practice because they favor the lower investment that will be required for reformulated gasoline. Certainly, in the short term, the 12–10 vote of the House of Representatives Subcommittee on Health and the Environment (October 11, 1989), against requiring the mass-production of alternative fueled vehicles by the automobile manufacturers, is unlikely to clarify matters.

Whether hydrogen will be available via a methanol intermediate, or as a direct energy vector, it will be necessary to store it for transportation applications. Storage of hydrogen in the vehicle is a technology that is either current, or is certainly feasible. In the F.R.G., BMW has shown that liquid hydrogen can be used, though problems associated with boil-off still remain. Even compressed hydrogen is feasible, if compressed natural gas is considered to be acceptable for an IC-engined vehicle. This can be illustrated by the increased fuel efficiency of a fuel cell vehicle compared with an IC engine, which more than compensates for the lower volumetric fuel efficiency (by a factor of three) of hydrogen in gaseous or liquid form compared with gaseous methane or liquid gasoline.

The measured urban driving cycle fuel economy of an Opel electric vehicle with regenerative braking at Texas A&M University, expressed as kW h (a.c.)/kW h_{th} of gasoline is 4.1. The real vehicle power requirements at the a.c. outlet are 0.15 kW h (a.c.)/km. Electrical efficiency of the vehicle (ratio of a.c. input to battery output) is 57%. With an average hydrogen fuel cell voltage of 0.65 V per cell, the fuel cell efficiency will be 44% based on the higher heating value of hydrogen. Thus, the ratio of energy use as hydrogen in the fuel cell to that of gasoline in the IC engine is 1:3.17. Thus, not only will the volume of hydrogen fuel be acceptable, so will the fuel cost per km.

It was gratifying to note that Elenco in Belgium advocates the use of minibuses with alkaline fuel cells and liquid hydrogen fuel. While the fuel cell still needs improvements in materials, it most of all requires improvements in engineering. Small systems with stainless steel Swagelock components clearly require redesign, as do large systems with separate boxes joined by long segments of piping covered with hand-wrapped insulation.

The paper presented by KTI stressed this point: cost will be lowered by cascading components and eliminating separate connections. Design must be tightened up, which should be possible as reliability increases and maintenance aspects of the system, which are facilitated by modular construction, become less critical. These design aspects are being carefully addressed by the Japanese developers in their field test programs. For stationary systems, weight is at first sight not very critical. For example, in the phosphoric acid system, the weight of active stack components (minus electrolyte) in the Westinghouse unit is now about 4.5 kg/m^2, 90% of which is the cross-flow graphite bipolar plate. This represents 3.75 kg/kW at current power densities. Thus, the active components represent about 1500 kg per stack. However, the weight of the latter, without the pressure vessel, is about 7.2 tonnes. Since weight is synonymous with cost, it would seem that lightweight designs must be evaluated in the future. In fact, the impression is that today's fuel cell designs are largely breadboards, made for accessibility rather than for optimum cost. The stationary systems (phosphoric acid, molten carbonate, solid oxide) all require simplification, lower weight, and thus lower cost. As an exaggerated example, the 3 kW solid oxide units contain approximately 25 kg of tubular cells, about 80% of which is the weight of the support tubes. However, the weight of the total system is 1300 kg. The phosphoric acid system still requires performance improvements to compete with the combined cycle. In contrast, the molten carbonate system can reach much higher efficiencies than this competitive system, but it still requires some materials improvements, for example, at the cathode. It also requires a cost-effective bipolar plate with cost-effective aluminizing for corrosion prevention. Similarly, the solid oxide system requires weight reduction and cost-effective manufacture.

For mobile power fuel cell systems, cost must be greatly reduced to levels that are perhaps 10% of those for stationary applications. Specific power must also be increased. Again, both of these imply weight reduction of the repeat parts of the cell stack. For a PEM system operating on air at ambient pressure, a maximum current density of about 4 kA/m^2 can now be obtained at 0.6 V. This represents 2.4 kW/m^2, or 480 kW/m^3 at a standard stacking pitch of 5 mm for cross-flow cells of standard configuration. Using present 'standard' components, this corresponds to 1000 kg/m^2, or 2.1 kg/kW for the basic d.c. module repeat components. However, it should be possible to improve the engineering design still further by improving the engineering design of the stack. Typically, a large graphite bipolar plate intended for a utility phosphoric acid system has 1.5 mm square gas distribution grooves in a cross-flow configuration in a plate with 0.5 mm web. A smaller plate for a mobile PEM system may have grooves only 1 mm square, allowing a stack specific weight reduction of about 20%, to give 4.1 kg/m^3 and 1.7 kg/kW, at the increased specific power per unit volume of 600 kW/m^3, resulting from the increased stacking density (4 mm cell pitch). If internal manifolding and a co- or counter-flow arrangement are used, the structure of the bipolar plate can be made much lighter and thinner. In such a case, the grooves are parallel on each side, and if the pitch on one side is displaced 90° out-of-phase with regard to the other, a thin folded or undulated plate can supply the gas channels, as in the Alsthom design of the late 1960s and early 1970s. As in this bipolar plate, a graphite–plastic composite may be suitable, though other possibilities exist. If this should be the case, a plate weighing 1.5 kg/m² is possible, giving a total component weight of 1.8 kg/m², including electrodes, electrolyte layer, and a wicking arrangement for water mass control, for example, by the use of thin conducting graphite felts. A cell pitch of 2 mm should therefore be possible, allowing a power density per unit volume of 1200 kW/m³, with a specific weight of 0.75 kg/kW.

Attempting to achieve this goal has nothing to do with electrochemistry, and everything to do with good engineering design. Electrochemically speaking, it is based on the state-of-the art current densities and power densities for the latest PEM cells operating under atmospheric pressure conditions at Los Alamos National Laboratory and at Texas A&M University. Naturally, such ambitious performance will require innovative engineering design in other associated components and subsystems, including cooling and gas flow distribution and control with allowable geometry for pressure drop. It also requires innovative lightweight end-plate design, with elimination of the standard tie-bolts and their replacement by, for example, a sealed tension system incorporating composite structures. For the ambient pressure PEM system, it is perhaps possible to achieve 1.125 kg/kW for the complete stack, with its hydrogen fuel and air supply and cooling systems.

These energy densities may seem to be unrealistic, until one considers that a projected alkaline fuel cell (International Fuel Cells) that will operate at 80 kA/m² on pure hydrogen and oxygen at 13.6 atm is now a real possibility. The projected weight for this system per unit area is 1.8 kg/m². For the cell stack alone, including an advanced cooling system and its auxiliaries and the pressure vessel, 7 kW/kg is projected. For a complete system, physically similar to that in the space shuttle, 300 kW is expected in a 90 kg unit. Such power densities are of the same order as those from military gas turbines, and they largely exceed the possibilities of normal internal combustion engines. To put them in another more historical perspective, the mid-60s alkaline fuel cell used in the Apollo program weighed 115 kg and produced 1.5 kW, whereas the three-stack version of the mid-70s space shuttle orbiter system produced 18 kW (limited by cooling) for the same system weight. Both of the above operated on cryogenic hydrogen and oxygen at 4 atm pressure.

A 'terrestrialized' atmospheric pressure version of this generator, operating on hydrogen and CO_2-scrubbed air at 70 °C (rather than 150 °C) could use inexpensive materials (plastics) and non-noble catalysts (at least at the cathode). With a pyrolyzed cobalt TAA-carbon catalyst (or similar compound), 4.5 kA/cm² at 0.65 V should be achievable, giving a power output of 60 kW from a stack weighing 40 kg, occupying 41 l. The complete system might weigh 90 kg, so that its weight and volume will be compatible with a family car. In mass production, an alkaline system could be inexpensive (perhaps only a few dollars per kW for the stack, based on an average materials cost of $2 - 3/kg or $4 - 6/kW, of which only 5% would be cobalt).

Similarly, there is no reason why a PEM should be expensive in the future. The present cost of Nafion® is over $3000/kg. It is made from the same starting material (tetrafluoroethylene) as Teflon®, which costs about $40/kg. The differential is disproportionate, but the cost of Nafion® is artificially inflated by its value to the chloro-alkali industry. There is no real reason why a PEM fuel cell stack, 10% of whose weight would be fluorocarbon acid polymer, could not also have an average materials cost that would ultimately not be very different from that of an alkaline stack. Even platinum electrocatalyst cost can be (marginally) acceptable: state-of-the art loadings of 1 and 2 g/m^2 can now be achieved at the anode and cathode respectively, based on work at Los Alamos National Laboratory and Texas A&M University. This represents 2.4 troy oz (75 g) for a 60 kW unit, $i.e.$, $22/kW. However, unless a major breakthrough in acid–electrolyte catalysis takes place, the use of platinum catalyst will be impractical: at the present catalyst loading, only 1.3 million cars per year can be supplied from the entire world production of platinum*. The present world vehicle population, with a 10-year turnover time, is 500 million, and it is projected to be 2 billion by the year 2010. Thus, a platinum-based fuel cell system could not begin to make an impact on the energy use of future vehicles and on their exhaust emissions.

Based on present knowledge, the best candidates for mobile applications will be the alkaline system and, more problematically, a monolithic solid oxide system, if it can be made. The difficulties involved in achieving this goal were carefully and convincingly summarized at this Symposium in the paper from Combustion Engineering. A definite advantage of the solid oxide cell is that its waste heat can be used in a thermal bottoming cycle, $e.g.$, by pressurizing the system, which will allow an increase in power density and perhaps in efficiency, thus further reducing cost. The power density of the low temperature systems will benefit greatly by pressurizing, but then extra electrical energy must be used to supply the high pressure air, unless some innovative method of compression can be developed, for example using the energy available in liquid hydrogen. For example, if this is stored under 70 atm pressure, it could pressurize ten times its volume of air (a typical requirement) when let down to 5 atm. While many authors suggest the use of pressurized PEM cells, the electrical requirement to provide pressurization may be the equivalent of 0.15 V, which would

---

*Editor's note: although a fuel cell powered car will not require a precious metal catalytic exhaust convertor; hence this application represents a switch in the location of platinum catalyst rather than a wholly new requirement. This aspect of supply will need to be confronted in either event, although fuel cell platinum catalyst levels (where required) would be two orders of magnitude higher than those for catalytic mufflers based on state-of-technology loadings.

seriously degrade overall system efficiency if it had to be provided electrically.

For the future, hydrogen must be given pre-eminence as the fuel cell fuel par excellence. While methanol may be the preferred future fuel for vehicles, this does not necessarily mean that a fuel cell vehicle should carry around a bulky load-following chemical plant. As already noted, a better solution would probably be to place a reformer at the service station. To 'fill the tank' of a hydrogen fuel cell car might typically require 150 kW h_{th} of hydrogen to give 450 km range. A gasoline pump can service about 10 vehicles per hour, 12 hours per day. Thus, a typical six-pump station would require a 4.5 MW reformer, operating continuously with hydrogen storage. This does not seem to be an impossible requirement.

Finally, the key to the future is the development of new materials. Not everything has been invented yet: new materials may allow a broadening of the scope of fuel cells, and allow their wider use. New solid electrolytes are part of this opportunity, and it is gratifying to know that the EEC has a far-sighted program in this area. The risk is high, but the opportunities and potential spin-off are great, for example, in the field of catalysis and super-conductivity. New lines for research exist in the area of new materials, which will serve to educate the electrochemists and electrochemical engineers which will be required for the future world of the electrochemical engine.

AUTHOR INDEX: GROVE ANNIVERSARY FUEL CELL SYMPOSIUM

SUBJECT INDEX: GROVE ANNIVERSARY FUEL CELL SYMPOSIUM